THE AMERICAN INSTITUTE OF ARCHITECTS

美国建筑师协会
2010~2012 获奖作品集
AIA 2010~2012 DESIGNS FOR THE NEW DECADE

美国建筑师协会/编 常文心/译

辽宁科学技术出版社

图书在版编目（CIP）数据

美国建筑师协会2010～2012获奖作品集 / 美国建筑师协会编；常文心译. -- 沈阳：辽宁科学技术出版社，2012.5
 ISBN 978-7-5381-7400-7

 Ⅰ．①美　Ⅱ．①美　②常　③　Ⅲ．①建筑设计－作品集－美国－2010～2012　Ⅳ．①TU206

中国版本图书馆CIP数据核字(2012)第035361号

出版发行：辽宁科学技术出版社
　　　　　（地址：沈阳市和平区十一纬路29号　邮编：110003）
印　刷　者：利丰雅高印刷（深圳）有限公司
经　销　者：各地新华书店
幅面尺寸：240mm×310mm
印　　张：55.25
插　　页：4
字　　数：100千字
印　　数：1～2000
出版时间：2012年 5 月第 1 版
印刷时间：2012年 5 月第 1 次印刷
责任编辑：陈慈良　王晨晖　常文心
封面设计：杨春玲
版式设计：杨春玲
责任校对：周　文
书　　号：ISBN 978-7-5381-7400-7
定　　价：380.00元

联系电话：024-23284360
邮购热线：024-23284502
E-mail: lnkjc@126.com
http://www.lnkj.com.cn
本书网址：www.lnkj.cn/uri.sh/7400

007 Foreword
前言

008 A History of the AIA Honor Awards Program
美国建筑师协会荣誉奖项目历史

010 Introduction
简介

012 2010 Institute Honor Awards Jury
2010 年美国建筑师协会荣誉奖评委

156 2011 Institute Honor Awards Jury
2011 年美国建筑师协会荣誉奖评委

304 2012 Institute Honor Awards Jury
2012 年美国建筑师协会荣誉奖评委

438 Index
索引

2010

ARCHITECTURE
建筑

016 Alice Tully Hall
爱丽丝杜莉厅

020 Austin E. Knowlton School of Architecture
奥斯丁·E·诺尔顿建筑学院

028 Beauvoir
美景

030 Brochstein Pavilion and Central Quad: Rice University
莱斯大学布洛奇斯坦馆和中央庭院

034 Camino Nuevo High School
卡米诺·尼沃高中

038 Campus Restaurant and Event Space
园区餐厅和活动空间

044 Macallen Building
麦卡伦大厦

048 Outpost
前哨

054 Serta International Center
舒达国际中心

060 Skirkanich Hall
斯科尔卡尼奇厅

068 Step Up on 5th
登上第五层

072 TKTS Booth and the Redevelopment of Father Duffy Square
TKTS 售票亭和杜菲神父广场重建工程

076 Urban Outfitters Corporate Campus
城客服饰公司园区

082 Yale University Art Gallery, Kahn Building Renovation
耶鲁大学美术馆——卡恩楼翻新

INTERIOR ARCHITECTURE
室内设计

086 CHANEL Robertson Boulevard
香奈儿罗伯森大道店

092 Craftsteak
工艺牛排店

096 Data
达塔公司

100 Exeter Schools Multipurpose Space
埃克塞特学校多功能空间

104 Historic Central Park West Residence
中央公园西部住宅

108 The Cathedral of Christ the Light
耶稣光明大教堂

114 Vera Wang Boutique
王薇薇精品店

REGIONAL & URBAN DESIGN
区域和城市规划

118 A Civic Vision for the Central Delaware River
德拉瓦河中游市政前景

124 Connections: MacArthur Park District Master Plan
连接：麦克阿瑟公园区整体规划

128 Greenwich South Strategic Framework
格林威治南区战略框架

132 Monumental Core Framework Plan
纪念碑核心区框架规划

138 Ryerson University Master Plan
怀尔逊大学总体规划

142 Savannah East Riverfront Extension
萨凡纳东部河岸区域扩建

148 The U.S. House Office Buildings Facilities Plan and Preliminary South Capitol Area Plan
美国众议院办公设施规划和南部国会区初步规划

TWENTY-FIVE YEAR AWARD
25 年

152 King Abdul Aziz International Airport – Hajj Terminal
阿卜杜勒阿齐兹国王国际机场朝觐航站楼

2011

ARCHITECTURE
建筑

160 AT&T Performing Arts Center Dee and Charles Wyly Theater
AT&T表演艺术中心迪和查尔斯·威利剧院

166 Ford Assembly Building
福特装配楼

172 Horizontal Skyscraper Vanke Center
水平摩天楼：万科中心

180 New Acropolis Museum
新卫城博物馆

184 North Carolina Museum of Art
北卡罗来纳艺术博物馆

188 One Jackson Square
杰克逊一号广场

194 San Francisco Museum of Modern Art Rooftop Garden
旧金山现代艺术博物馆屋顶花园

198 The Barnard College Diana Center
巴纳德学院戴安娜中心

204 University of Michigan Museum of Art
密歇根大学艺术博物馆

210 U.S. Land Port of Entry
美国陆上入境口岸

INTERIOR ARCHITECTURE
室内设计

216 Alchemist
炼金术师服装店

220 Armstrong Oil and Gas
阿姆斯特朗油气公司

226 Conga Room
康加舞厅

234 FIDM San Diego Campus
时装设计商业学院圣地亚哥校区

240 Moving Picture Company
移动图形公司

246 Registrar Recorder County Clerk Elections Operations Center
登记员／县书记官选举运营中心

250 The Academy of Music
音乐学院剧院

254 The Power House, Restoration/Renovation
电力站修复翻新

258 Vancouver Convention Center West
温哥华会展中心西区

264 Washington Square Park Dental
华盛顿广场公园牙科诊所

270 John E. Jaqua Center for Student Athletes
约翰·E·雅克大学生运动员中心

REGIONAL & URBAN DESIGN
区域和城市规划

276 Beijing CBD East Expansion
北京中央商务区东扩规划

278 Chicago Central Area DeCarbonization Plan
芝加哥中心区脱碳规划

284 "Community | City: Between Building and Landscape Affordable Sustainable Infill for Smoketown, Kentucky"
"社区｜城市：烟镇建筑与景观之间的可持续填充设计"

290 Gowanus Canal Sponge Park
郭瓦纳斯运河海绵公园

292 "Low Impact Development: A Design Manual for Urban Areas"
《低影响开发：城区规划手册》

298 Townscaping an Automobile-Oriented Fabric
以机动车为主的城市网格中的城市景观规划

TWENTY-FIVE YEAR AWARD
25年

302 John Hancock Tower
约翰·汉考克大厦

2012

ARCHITECTURE
建筑

- 308 **8 House**
  8 字住宅
- 320 **41 Cooper Square**
  库伯广场 41 号
- 324 **The Gates and Hillman Centers for Computer Science**
  盖茨和希尔曼计算机科学中心
- 330 **Ghost Architectural Laboratory**
  幽灵建筑实验室
- 334 **LumenHAUS**
  流明屋
- 338 **Pittman Dowell Residence**
  皮特曼·道威尔住宅
- 342 **Poetry Foundation**
  诗歌基金会
- 348 **Ruth Lilly Visitors Pavilion**
  卢斯莉莉游客亭
- 352 **The Standard, New York**
  纽约标准酒店

INTERIOR ARCHITECTURE
室内设计

- 360 **ARTifacts**
  工艺工作室
- 364 **Children's Institute, Inc. Otis Booth Campus**
  儿童学院公司奥的斯园区
- 370 **David Rubenstein Atrium at Lincoln Center**
  林肯中心大卫·鲁宾斯坦中庭
- 374 **HyundaiCard Air Lounge**
  现代信用卡机场休息室
- 380 **The Integral House**
  整合住宅
- 384 **Joukowsky Institute for Archaeology & the Ancient World**
  儒科夫斯基考古和远古世纪学院
- 388 **Memory Temple**
  记忆殿
- 392 **Prairie Management Group**
  大草原管理集团
- 396 **Record House Revisited**
  实录房重建
- 400 **The Wright at the Guggenheim Museum**
  古根海姆博物馆莱特餐厅

REGIONAL & URBAN DESIGN
区域和城市规划

- 404 **Fayetteville 2030: Transit City Scenario**
  费耶特维尔 2030：城市交通情景规划
- 408 **Grangegorman Urban Quarter Master Plan**
  格兰格曼城区总体规划
- 412 **Jordan Dead Sea Development Zone Master Plan**
  约旦死海开发区总体规划
- 418 **Master Plan for the Central Delaware**
  德拉瓦河中段总体规划
- 420 **Miami Beach City Center Redevelopment Project**
  迈阿密海滩城市中心再开发项目
- 424 **Portland Mall Revitalization**
  波特兰林荫大道复兴工程
- 428 **Reinventing the Crescent: Riverfront Development Plan**
  改造新月区：河畔开发规划
- 430 **SandRidge Energy Commons**
  沙波能源公司绿地规划

TWENTY-FIVE YEAR AWARD
25 年

- 434 **Gehry Residence**
  盖里住宅

FOREWORD
前言

Since 1949, the American Institute of Architects has encouraged design excellence through its Institute Honors Awards program. Hundreds of projects, firms, and architects have been honored for their ingenuity in a range of categories. Yet, all of them have achieved the same high standard of rigor, clarity, and spirit.

The projects that were premiated in 2010, 2011, and 2012 are no different and speak to the quality of global architectural production in the first decade of the new century. They also speak to a few interrelated themes: regional adaptation to climate, soil, infrastructure, and culture; a sense of context and place that drives the project, conceptually; a strong desire on the part of the design architect to balance the building typology with local tradition and community.

Of course, there are some global themes that connect the work here, however diverse it may be: the urgency of finding sustainable solutions, the economy of systems-thinking in design, and the urbanity of public spaces to bring us all together.

As you learn more about these projects, consider the whole to be a survey of talent and a good overview of the quality that an AIA member brings to the project at hand. Naturally, the scale of work in countries like China is much larger than in the United States. But, the finer grain work that happens in the United States represents a living laboratory – constantly examining, re-examining, building, and evolving. I think you'll understand more fully what that laboratory looks like as you mine the pages of this book.

Robert A. Ivy, FAIA
Executive Vice President/Chief Executive Officer
The American Institute of Architects

自1949年起，美国建筑师协会就通过美国建筑师协会荣誉奖项目来鼓励优秀的设计。成百上千的项目、公司和建筑师因他们的出色表现被授予了不同类别的奖项。但是，他们都达到了同等高度的精确、明晰和精神。

2010、2011和2012三年的获奖项目同样如此，彰显了本世纪首个十年全球建筑的质量。同时，他们还代表了一系列相关联的主题：针对气候、土壤、基础设施和文化的区域性改革；驱动项目的情境感和地方感；设计建筑师对平衡建筑类型和当地传统及社群的强烈愿望。

当然，一些全球化主题连接了这些项目，尽管主题可能各有不同：找到可持续解决方案的紧迫性、设计中系统思考的经济性以及将我们带到一起的公共空间的城市风格。

当你深入了解这些项目时，请将它们看作是才智的反馈和美国建筑师协会成员们所设计项目的整体质量总览。很自然地，中国的项目规模要远大于美国。但是美国项目更好的质感代表了一个鲜活的实验室——不断地检查、再检查、建造和进化。阅读本书，我相信你会更完整地理解这个实验室。

——罗伯特·A·艾维（美国建筑师协会会员）
执行副主席/首席执行官
美国建筑师协会

A HISTORY OF THE AIA HONOR AWARDS PROGRAM[i]

美国建筑师协会荣誉奖项目历史[i]

The Institute Honor Awards program of The American Institute of Architects encourages distinguished design by recognizing it and celebrating its architects. Yet this is only part of the meaning of this awards program, as a study of the recipients from 2010 through 2012 will reveal. It is a striking representation of where architecture stands today and where it promises to lead.

The importance of this portfolio is not limited to the architect, whose natural preoccupation is design, nor is it limited to the critic. It is not limited to engineering or construction, which are both guided by and informed by the design process.

Architecture is about making and re-making of the physical environment in which we live, work, play, and learn. It's about designing at all scales and inspiring others to think from the spoon to the city, to paraphrase the architect Ernesto Rogers. It's about the synthesis of past lessons and today's aspirations for the benefit of future generations.

Since its inception in 1857, the AIA has helped advance all of these ideas for an architecture profession in service to society. As it neared its centennial anniversary, the Institute's leadership realized that encouraging good design among its members was not enough; it must also encourage public notice and acclaim for good design if the profession was to continue making a difference in the built environment.

In 1948, at the convention held in Grand Rapids, Michigan, the AIA Board of Directors formed a committee to bring before the Institute a concrete proposal for an honor awards program centered on current work.

A year later, in 1949, the AIA launched its Honor Awards for Current Work program.

There are two main ways of narrowing the field of entrants in an awards program – by membership type and by awards category. The AIA diligently experimented with both. At first, organizers solicited only corporate members of the Institute to submit. By 1950, they lifted that restriction and accepted entries from anyone legally entitled to call themselves an architect. Over the next few years, the program was governed by categories rather than membership type. Those categories (based on building type) rotated annually based on the AIA Board or Executive Committee's recommendations based on what was most desirable for the upcoming annual convention: schools, churches, hospitals, residences, commercial buildings, and so on.

But, as Edmund Purves, FAIA, the consulting director for the AIA's awards program in 1962 acknowledged, even this arrangement had its limitations. "The categories were poorly balanced one with another," noted Purves, "and there were one or two rather sad occasions when the category failed to achieve a single mention. Unexpectedly the use of categories became a rather invidious restriction."[ii] The Honor Awards were opened to any type of architecture.

One of the first Honor Award recipients was Skidmore Owings and Merrill for Lever House in New York. Completed in 1952, the building is now regarded as the quintessence of International Style office buildings, certainly owing to its appearance, but also to its commitment to a civic ideal. While an office building's "public plaza" is commonplace today, Bunshaft and de Blois set a generous standard here that remains, in many ways, peerless. Design architects Gordon Bunshaft, FAIA, and Natalie de Blois, FAIA, incorporated New York's first curtain wall system, making Lever House an engineering touchstone as well as an icon of Modernism.

Over the years, additional sub-categories have come and gone, such as the Bartlett Award to recognize Barrier-Free Architecture in the early 1970s, before handicapped access was a legal requirement. One of the notable recipients of the Barlett Award was Louis Kahn, FAIA, for his Kimball Museum of Art in Fort Worth, Texas in 1975. The Kimball's 16 cycloid vaults, spread out over one level make it a model of accessibility. It's also a model of innovation and collaboration. From the skylight baffles (which diffuse the sun evenly to produce a silvery glow) to the surrounding landscape, Kahn worked closely with the engineer August Komendant, the landscape architect George Patton, and the lighting designer Richard Kelly to create what is regarded as a critical turning point in museum design.

In some cases, buildings have repeatedly surfaced as touchstones of good design for generations of architects. The Santa Monica home of Frank Gehry, FAIA, which he renovated (for the first time) in 1978, received an honor award in 1980. In 2012, it emerged in the award rolls again to receive the AIA 25 Year Award. The irony, of course, is not lost – for such a modest highly personal project, constructed with unglamorous materials on a shoestring budget, to have lasting impact and widespread significance is a testament to Gehry's particular talent.

Architecture's ability to adapt, as evinced by Gehry's home, was not lost on awards organizers, either. A separate division for "Extended Use" was added in the early-1980s to recognize the reuse of historic structures and restoration. In 1994, categories to recognize Interior Architecture and Regional and Urban Design were also added. Inclusiveness aside, the focus of the Institute Honor Awards has remained singular – to recognize the best examples of contemporary architecture.

In the pages of this book, the AIA presents recipients in the Architecture, Interior Architecture, and Regional and Urban Design categories within the Institute Honors Program. The breadth of architectural production is on full display here in a series of fine projects completed in the first decade of the 21st century. These projects are standard bearers for good design, in keeping with the mission of the AIA's awards programs. But, each one also represents the fruits of a collaborative and rigorous design process. In doing so, they transcend our expectations for what architecture can achieve.

美国建筑师协会荣誉奖项目通过对优秀设计进行认证并表彰其建筑师来鼓励更多的优秀设计。但是，通过对2010年到2012年之间获奖者的研究，你会发现，这仅是该奖项意义的一部分。它代表了当今建筑的水平以及建筑未来的走向。

荣誉奖的重要价值不仅局限于建筑师（他们的职业就是设计）、评论家、工程或建设（这二者通过设计流程而得到指导）。

建筑师对我们生活、工作、娱乐和学习的环境的创造和再创造过程。它包含一切层面的设计，激发人们从城市的角落一直思考到重新诠释建筑师厄耐斯特·罗杰斯的作品。它是对过去经验的整合，也是对下一代的启发。

自1857年创建以来，美国建筑师协会就一直帮助促进建筑对社会的服务。在协会进行百年纪念的时候，协会的领导人认识到了单单鼓励其成员的优秀设计已经不够，协会必须还鼓励公众对好设计的认证。协会的任务是持续让建成环境与众不同。

1948年，在密歇根州大急流城所举办的会议中，美国建筑师协会董事会成立了一个委员会，具体地提出了当前作品荣誉奖奖励方案。

一年之后的1949年，美国建筑师协会启动了当前作品荣誉奖项目。

有两种主要方式来缩小奖项的参赛者——以会员为限制，或者以奖项分类来限制。美国建筑师协会对两种方式都进行了试验。首先，组织者只向协会的企业成员寻求参赛作品。1950年起，他们降低了限制，开始接受所有合法建筑师的作品。在之后的几年中，荣誉奖项目通过奖项分类来进行限制。这些分类（以建筑类型为基础）根据美国建筑师协会董事会或执行委员会的推荐而逐年变化，以来年的建筑师大会主题为基础，例如：学校、教堂、医院、住宅、商业建筑等。

但是，正如1962年美国建筑师协会荣誉奖项目的咨询总监艾德蒙·普尔维斯所说，这种安排也有局限。"这些分类十分不平衡"，普尔维斯称，"有时候某个分类甚至无法选出一个提名作品。分类的运用相当不公平。"ⁱⁱ 荣誉奖应该对所有建筑类型开放。

第一批荣誉奖获奖作品之一是SOM事务所设计的纽约利华大厦。建筑完成于1952年，被认为是国际风格办公楼的典范，这主要归功于它的外观和它对市政典范的贡献。现在，办公楼前的"公共广场"已经司空见惯，戈登·邦夏和纳塔利·布洛瓦在这里建立了标准。从很多方面上讲，这座建筑都无与伦比。设计是建筑师邦夏和布洛瓦引进了纽约的第一个幕墙系统，让利华大厦成为了工程试金石和现代主义的标志。

多年以来，额外的附属分类变化不定。例如：20世纪70年代早期的巴特利特奖用于认证无障碍建筑，当时无障碍入口还没被列入法规要求。巴特利特奖的得主之一金博尔设计的16摆线拱顶。建筑的各个部分都设在同一平面，成为了可达性的典范。它还是创新和合作的典范。从天窗挡板（挡板能够漫射阳光，打造银色的光晕）到周边景观，卡恩与工程师奥古斯特·科曼戴特、景观设计师乔治·巴顿以及灯光设计师理查德·凯利紧密合作，创造了博物馆设计的转折点项目。

在某些情况下，建筑不断地成为建筑师们的好设计试金石。弗兰克·盖里的圣塔莫尼卡住宅建于1978年，并于1980年获得了建筑荣誉奖。在2012年，它重新进入到了获奖名单中，获得了美国建筑师协会25年奖。这样一座以单调材料和紧缩预算建造的低调的私人项目竟然拥有持续的影响力和广泛的意义，体现了盖里的独特天赋。

正如盖里住宅所显示，建筑的适应能力不会不被奖项的组织者所注意。20世纪80年代，荣誉奖添加了"扩建"类别，拥有认证历史建筑的重新利用和翻修项目。1994年，荣誉奖还添加了室内设计和区域与城市规划两个奖项。尽管荣誉奖具有很强的包容性，它的焦点始终如一——认证现代建筑的最佳例证。

本书呈现了美国建筑师协会建筑、室内设计和区域与城市规划荣誉奖的获奖作品。本书以21世纪首个十年的一系列优秀项目充分展现了建筑作品的广度。这些项目好设计的倡导者，与美国建筑师协会的任务相一致。但是，每个项目都是合作与精密的设计流程的结果。它们超出了我们对建筑的期待。

---

i. Adapted from Philip Will, Jr.'s forward and Edmund R. Purves, FAIA's, introduction to Mid-Century Architecture in America: Honor Awards of the American Institute of Architects, 1949-1961. Baltimore: Johns Hopkins University Press (1962): 5-7, 29-30.
ii. Purves, Edmund R., FAIA, "The AIA 'Honor Awards for Current Work' and its Juries," in Wolf Von Eckardt (ed.), Mid-Century Architecture in America: Honor Awards of the American Institute of Architects, 1949-1961. Baltimore: Johns Hopkins University Press (1962): 30.

i. 选自 Philip Will, Jr.'s forward and Edmund R. Purves, FAIA's, introduction to Mid-Century Architecture in America: Honor Awards of the American Institute of Architects, 1949–1961. Baltimore: Johns Hopkins University Press (1962): 5–7, 29–30.
ii. 选自 Purves, Edmund R., FAIA,"The AIA 'Honor Awards for Current Work' and its Juries," in Wolf Von Eckardt (ed.), Mid-Century Architecture in America: Honor Awards of the American Institute of Architects, 1949–1961. Baltimore: Johns Hopkins University Press (1962): 30.

INTRODUCTION

简介

Peer recognition drives professional excellence. That's true in every profession and in all creative endeavors — and architecture uniquely draws from both of those worlds. It is bound by the ethical codes, rigorous training, and licensure process like law or medicine; it is also the art of building in service to the global population. Although the demographics of its membership have shifted over time, the American Institute of Architects has always represented the professionals who practice this art. The AIA's Institute Honor Awards program annually chronicles excellence in thinking, collaboration, and ultimately, design.

In the pages that follow, the AIA recognizes three years worth of award recipients across four categories: architecture, interior architecture, regional and urban design, and its esteemed Twenty-five Year Award. On one hand, these projects from 2010, 2011, and 2012 constitute a snapshot of architectural production — a moment in time and a record of "excellence" in that moment. On the other hand, it is a snapshot that speaks more broadly about evolving design trends — what has come before, in the first decade of the 21st century, and what we might expect more of in the following decade. As one juror noted, these projects were selected from a portfolio of exceptional work that survived rigorous critique, discussion, site visits, and extensive deliberations. Although all of the entries were conceived with passion and commitment, those that garnered an award uniquely demonstrated a single, compelling idea and purpose. In essence, they constitute design excellence both in this moment and despite the vicissitudes of an age-old art form.

Across the four categories contained in this book, there are some constants worth noting: projects are considered individually within their categories, rather than relatively to other projects in the same category. Projects must also be more than simply unique or interesting; they must be attentive to their contexts and the environmental issues that deeply affect us all. However, each entry in the Institute Honor Awards program is judged against the degree to which it has met the category's individual requirements.

Additionally, for each of the following categories, all submissions must include the project's percentage of energy reduction and energy consumption (per square foot) as defined by the U.S. Environmental Protection Agency's (EPA) Energy Star Target Finder Tool and/or documentation of specific material choices to address the needs for indoor environmental quality and diversion of materials from the waste stream. This is in recognition of the AIA Sustainable Architectural Practice Position Statement, which sets a goal of at least 50-percent reduction of fossil-fuel energy use by 2010 and carbon neutrality by 2030.

For the Institute Honor Award for Architecture, projects must exhibit design achievement that demonstrates exemplary skill and creativity in resolution and integration of formal, functional, and technical requirements, including ecological stewardship and social responsibility that acknowledges and advances social agendas. Projects should reflect a strong sense of place, of ecology, of history, or of purpose as an integral part of the demonstrated design excellence.

In addition to design achievement, projects may be exemplary in the following subcategories: technical advancement, which includes engineering achievements (structural, mechanical, transportation, computer, etc.) as well as innovative use of materials; and/or preservation/restoration, including demonstration of exemplary skill, sensitivity, and thoroughness in preservation, restoration, or alternative use of existing buildings regardless of their original architectural significance.

The Institute Honor Awards for Interior Architecture acknowledges the excellence of building interiors created by architects licensed in the United States. Program organizers intend to draw attention to the full range of completed interior architecture: entries may be large or small in scope; they may involve renovation or adaptive use; they may also represent new construction. Submissions in such areas of residential, institutional, commercial, corporate, retail, hospitality, or other focus are welcome. And all entries are judged on merit regardless of scale or budget.

The purpose of the Institute Honor Awards for Regional and Urban Design is to recognize distinguished achievements that involve the expanding role of the architect in urban design, city planning, and community development. The awards seek to identify projects and programs that involve public participation and contribute to the quality of the urban environment.

Owners, individual practitioners, private design firms, public agencies, civic organizations, and public interest groups may submit nominations for projects and programs in which they were involved. Applicants do not need to be architects or members of the AIA, but an architect licensed in the United States must be the author of the project.

Submissions may include urban design projects, planning programs, civic improvements, environmental programs, and redevelopment projects. Since many urban design projects are never "completed" in the traditional sense, "incomplete" projects or ongoing programs may be recognized if a significant portion has been completed, implemented, or adopted by a local jurisdiction.

Design achievement can be evidenced by the exploration of new approaches to ecological planning, urban form, or sensitive reinforcement of successful historical development patterns. Entries should address ecological issues by describing (preferably with graphics) how the design captures, collects, stores, and distributes resident renewable resources and energies. Entries may also exhibit improvements in the quality of life, the environment, and/or the technical advancement of urban systems.

For the Twenty-five Year Award, projects that receive this recognition have stood the test of time, having been completed between 25 and 30 years ago. They may be built in the United States or in some other country, but they must have been designed by an architect licensed in the United States. The award is open to architectural projects of all classifications and may be one building or a related group of buildings forming a single project.

The project must be standing in a substantially completed form and in good condition and it must still carry out the original design. Change of use is permitted when it has not basically altered original intent. The project must have excellence in function – in the distinguished execution of its original program and in the creative aspects of its statement by today's standards. Building and site together should be examined.

As you review these projects, remember that they represent countless hours of work by thousands of individuals working across time zones and, in all cases, over many years. They represent what a jury of architect peers has deemed worthy of recognition. It is recognition of accomplishment, to be sure, but also of the promise that tomorrow's architecture will have taken a cue from the best of today's excellent work.

William Richards,
The American Institute of Architects

同行的认可能促进人们追求专业卓越。这适用于所有行业和创造性活动——而建筑正好占了二者。一方面，建筑受道德标准、严格训练和行业许可流程（如法律和医药行业一样）的约束；另一方面，建筑艺术服务于全人类。尽管成员数量一直在变化，美国建筑师协会一直代表着从事这种艺术的专业人士。美国建筑师协会一年一度的荣誉奖项目记录了优秀的思想、合作及最重要的设计。本书呈现了美国建筑师协会三年以来在四个类别中的获奖作品：建筑、室内设计、区域与城市规划以及备受推崇的25年奖。一方面，这些来自2010年、2011年、2012年三年的项目构成了建筑作品的缩影——记录了当时的优秀设计和重要时刻。另一方面，它们还从广义上展现了建筑的进化趋势——21世纪第一个十年的设计以及未来我们能够期待的创新。正如一位评委所说：这些项目从严格的批评、讨论、现场考察和大量的协商中脱颖而出。尽管所有的参赛作品都充满了激情与努力，那些获奖作品展示了独具特色的理念和意图。这些项目从根本上展示了当下的优秀设计以及这种古老艺术形式的变迁。

本书中所收录的四类获奖作品有以下值得注意的地方：项目在其所在分类里是独立的，与其他项目没有联系。项目必须在独特或有趣之外有其他的优异之处；它们必须充分尊重周边文脉和环境问题。然而，每件学院奖项目的参赛作品都经过了其类别内独特要求的评判。

此外，在每个分类中，所有参赛作品都必须附加美国环保署能源之星标准评定的节能减排比例和能源消耗以及材料使用的详单（用于表明室内环境质量以及材料的废弃率）。这些都由美国建筑师协会可持续建筑实践状况报表的认证，它的目标是在2010年减少至少50%的化石能源使用并且在2030年达到碳平衡。

建筑荣誉奖的获奖项目必须展现出具有可借鉴性的技巧和在造型、功能和技术层面（包括生态管理和社会责任）上的创意解决方案。项目应该反映强烈的地方感、生态感、历史感，或者优秀设计的某些方面。

除了设计成就之外，项目还应在以下范畴内展现模范作用：技术进步（包括工程成就——结构、机械、交通、计算机等和材料的创意运用）及保护或修复（包括可借鉴的技巧、敏感性、全盘保护、修护或建筑全新的调整性使用）。

室内设计荣誉奖用于奖励美国注册建筑师所作出的优秀室内设计。项目组织者想要展示各种各样的室内设计：参赛作品可大可小，可以是翻新或重新利用，也可以是新工程。参赛作品涉及住宅、公共结构、商业、企业、零售、酒店或其他领域。评审过程中不会考虑规模或预算，只会考虑它们的优点。

区域和城市规划荣誉奖用于奖励建筑师在城市规划、城市设计及区域开发中的非凡成就。该奖项力求奖励在公众参与和环境质量中做出贡献的项目和规划。

项目所有人、独立从业者、私人设计公司、公共机构、市政组织以及公共利益群体都能以他们所参与的项目而获得荣誉奖提名。参赛人无需是建筑师或美国建筑师协会的成员，但是项目必须有美国注册建筑师的参与。

参赛项目可以是城市设计项目、规划项目、市政改造、环境项目或再开发项目。由于许多城市规划项目最终都没有"完成"，"未完成"或在建项目如果已经完成了重要的部分或者被区域当局所采纳，都可以参与到奖项角逐之中。

设计成就可以通过探索新的生态规划、城市模式或敏感性历史城区改造开发来实现。参赛作品应该通过描述（最好有图表）设计是如何捕捉、采集、存储并分配可再生资源和能源来表明自己的生态特征。项目还可以展示其对生活质量、环境以及城市系统的技术进步做出的贡献。

获得25年奖的项目应当经得住时间考验，建成时间在25到30年之间。它们可以在美国境内或其他任何国家，但必须由美国注册建筑师所设计。该奖项面向所有类别的建筑项目，可以是一座建筑，也可以是由建筑群组成的整体项目。

项目必须保持完整的造型和良好的状态，并且始终执行初始设计方案。只要不从根本上调整设计意图，建筑的用途是可以改变的。项目必须在功能上有杰出的表现——出色地执行原始项目规划并且以创新方式反映了当前的标准。建筑与场地将被一起检查。

在浏览这些项目时，请记得它们是由成千上万个个体经过无数个日夜的工作而建成的，有些项目甚至持续了多年。它们呈现了建筑师评委所认证的有价值设计。获奖项目当然获得了建筑成就的认可，同时，这些优秀的当代建筑作品也会为未来的建筑提供典范。

——威廉姆·理查德（美国建筑师协会）

2010 INSTITUTE HONOR
AWARDS FOR ARCHITECTURE
JURY
建筑荣誉奖评委

Richard L. Maimon, FAIA, Chair
KieranTimberlake Associates, LLP
理查德·L·迈蒙
美国建筑师协会，评委会主席
基兰－廷伯莱克建筑事务所

Jeanne Gang, FAIA
Studio/Gang Architects
珍妮·甘
美国建筑师协会会员
甘建筑事务所

Sam Grawe
Dwell/At Home in the Modern World
山姆·格拉维
生活在现代世界事务所

Jeffrey Lee, FAIA
Pearce Brinkley Cease & Lee P.A.
杰弗里·李
美国建筑师协会会员
皮尔斯·布林克利·西斯和李事务所

Justine N. Lewis
Georgia Institute of Technology/
American Institute of Architecture Student
Representative
贾斯汀·N·路易斯
乔治亚理工学院/美国建筑师协会学生代表

Miguel A. Rivera Agosto, AIA
Miró Rivera Architects
米格尔·A·里维拉·阿果斯托
美国建筑师协会
米罗·里维拉建筑事务所

Mark Simon, FAIA
Centerbrook Architects & Planners
马克·西蒙
美国建筑师协会会员
中央布鲁克建筑规划事务所

H. Ruth Todd, AIA
Page & Turnbull Architects
H·卢斯·托德
美国建筑师协会
佩吉和特恩布尔建筑事务所

William R. Turner, Jr., Associate AIA
Shears Adkins Architects
威廉姆·R·特纳二世
美国建筑师协会助理
希尔斯·阿德金斯建筑事务所

2010 Institute Honor Awards
2010年美国建筑师协会建筑/室内设计/
区域和城市规划荣誉奖评委

2010 INSTITUTE HONOR AWARDS FOR INTERIOR ARCHITECTURE JURY
室内设计荣誉奖评委

Daniel H. Wheeler, FAIA, Chair
Wheeler Kearns Architects, Inc.
丹尼尔·H·维勒
美国建筑师协会会员；评委会主席
维勒·吉恩斯建筑事务所

David H. Hart, FAIA
Utah Capitol Preservation Board
大卫·H·哈特
美国建筑师协会会员
犹他州议会大厦保护董事会

Audrey A. Matlock, AIA
Audrey Matlock Architect
奥德利·A·马特洛克
美国建筑师协会
奥德利·马特洛克事务所

Audrey Stokes O'Hagan, AIA
Audrey O'Hagan Architect
奥德利·斯托克斯·欧哈根
美国建筑师协会
奥德利·欧哈根建筑事务所

Clive R. Wilkinson, AIA, RIBA
Clive Wilkinson Architects
克里夫·R·威尔金斯
美国建筑师协会；英国皇家建筑师协会
克里夫·威尔金斯建筑事务所

2010 INSTITUTE HONOR AWARDS FOR REGIONAL AND URBAN DESIGN JURY
区域和城市规划荣誉奖评委

John F. Torti, FAIA Chair
Torti Gallas & Partners, Inc.
约翰·F·托尔蒂
美国建筑师协会会员；评委会主席
托尔蒂·格拉斯事务所

Lance Jay Brown, FAIA
Lance Jay Brown Architecture & Urban Design
兰斯·杰·布朗
美国建筑师协会会员
兰斯·杰·布朗建筑和城市规划事务所

Brenda Scheer, AIA
University of Utah
College of Architecture + Planning
布兰达·舒尔
美国建筑师协会
犹他大学
建筑规划学院

Edward K. Uhlir, FAIA
Uhlir Consulting, LLC
爱德华·K·乌利尔
美国建筑师协会会员
乌利尔咨询公司

Debby Wieneke
Habitat for Humanity of Benton County, Inc.
德比·维恩尼克
本顿郡仁爱之家公司

Richard L. Maimon, FAIA, LEED AP
2010 Chair,
Institute Honor Awards for Architecture
理查德·L·迈蒙
美国建筑师协会会员；美国绿色建筑协会认证专家
2010美国建筑师协会建筑荣誉奖评委会主席

© Ed Wheeler 埃德·维勒

Daniel H. Wheeler, FAIA
2010 Chair,
Institute Honor Awards for Interior Architecture
丹尼尔·维勒
美国建筑师协会会员
2010美国建筑师协会室内设计荣誉奖评委会主席

© Julie Wheeler 朱莉·维勒

Richard Maimon is a Principal at KieranTimberlake, an internationally recognized architecture firm noted for its integration of research and practice guided by a deep environmental ethic. He has been with KieranTimberlake for over twenty years, participating in the growth of the firm and deeply involved in the breadth of its work. He currently oversees a range of projects including the Embassy of the United States in London, UK, the Center City Building for the University of North Carolina at Charlotte, the Kimmel Center Master Plan, a housing prototype for the Make It Right Foundation in New Orleans, and the redesign of Dilworth Plaza in Philadelphia. He has been responsible for highly acclaimed projects including Melvin J. and Claire Levine Hall at the University of Pennsylvania, Atwater Commons at Middlebury College, F. Otto Haas Stage at the Arden Theater Company, and the Philadelphia Theatre Company's Suzanne Roberts Theater. Projects he has been responsible for have been published internationally and have received national design awards.

Mr. Maimon served as jury chair for the 2010 AIA Institute Honor Awards and the 2010 Twenty-Five Year Award. He is a frequent guest lecturer, with appearances at colleges and universities, AIA Chapters and national conferences including the AIA Convention in Boston, the North American Theater Engineering and Architecture Conference in New York City, and the United States Green Building Council Convention. He serves on the board of the Arden Theatre Company, a leading regional theater in Philadelphia, is on the Charter High School for Architecture and Design Business Advisory Council, and is active with other nonprofit organizations in Philadelphia.

Mr. Maimon earned a Bachelor of Architecture, magna cum laude, from Columbia University in 1985 and a Master of Architecture from Princeton University in 1989. He was awarded the Phi Beta Kappa Award from Columbia University in 1985. KieranTimberlake creates beautifully crafted, thoughtfully made designs which are holistically integrated to site, program and people. The firm is recognized for its research-based practice that focuses on new materials, processes, assemblies and products, receiving over one hundred design citations, including the 2008 Architecture Firm Award from the American Institute of Architects and the 2010 Cooper-Hewitt National Design Award.

理查德·迈蒙是基兰－廷伯莱克建筑事务所（一家国际知名建筑公司，以其对环保理念的研究和实践而著称）的总监。他在基兰－廷伯莱克建筑事务所奋斗了20余年，参与了公司的发展，在其中起到了举足轻重的作用。他最近所监督的项目包括：美国驻英国伦敦大使馆、北卡罗来纳大学的中心城市楼、齐默尔艺术中心总体规划、"正确行事"基金会在新奥尔良的住宅标准户型设计以及费城迪尔沃斯广场的重新设计。他所负责的知名项目包括：宾夕法尼亚大学的梅尔文·J和克莱尔·莱文厅、明德学院的亚特华德会堂、雅顿剧院公司的F·奥托·哈斯舞台以及费城剧院公司的苏珊娜·罗伯茨剧院。他所负责的项目已经在海内外发表出版，获得了许多国家设计大奖。

迈蒙先生是2010美国建筑师协会建筑荣誉奖和25年大奖的评委会主席。作为一名客座教授，他是各大院校、美国建筑师协会各地分会和国家建筑会议的常客。这些国家级会议包括：波士顿的美国建筑师协会大会、纽约的北美剧院工程建筑会议和美国绿色建筑协会大会等。同时，他还身兼雅顿剧院公司（费城的一家顶尖剧院）董事和查特高中建筑和设计顾问，并且活跃在费城其他非营利性机构中。

迈蒙先生于1985年在哥伦比亚大学以优秀成绩获得了建筑学学士学位，并于1989年在普林斯顿大学获得了建筑学硕士学位。1985年，他在哥伦比亚大学获得了美国大学优等生荣誉学会奖。

基兰－廷伯莱克建筑事务所创造工艺精美的设计，全面地适应场地、功能设置以及人的需求。公司的调研时间聚焦于新材料、新方法、新装配过程和新产品，获得了上百个设计荣誉，其中包括2008美国建筑师协会建筑公司大奖和2010库珀－休伊特国家设计奖。

Daniel H. Wheeler, FAIA, is principal of Wheeler Kearns Architects, and Professor of Architecture at the University of Illinois at Chicago. He has served as Interim Director for the UIC School of Architecture and the Graham Foundation for the Advanced Studies in the Fine Arts, and has been a collaborator with Auburn University's Rural Studio for the past ten years. A graduate of RISD, he worked in the early studio of Machado Silvetti in Boston. Prior to founding WKA in 1987, he was a Studio Head/Associate at Skidmore, Owings, and Merrill in Chicago.

丹尼尔·维勒（美国建筑师协会会员）是维勒·吉恩斯建筑事务所的总监，也是芝加哥伊利诺伊大学建筑学教授。他曾是芝加哥伊利诺伊大学建筑学院和格兰厄姆基金会美术进修部的临时主管，也曾在过去的10年间担任了奥本大学乡村研究室合作者。他毕业于罗德岛设计学院，曾在波士顿的马查多·希尔维蒂工作室工作。在1987年成立维勒·吉恩斯建筑事务所之前，他曾在芝加哥SOM公司担任工作室总监/助理。

John Francis Torti, FAIA, LEED AP
2010 Chair,
Institute Honor Awards for Regional & Urban Design
约翰·弗朗西斯·托尔蒂
美国建筑师协会会员；美国绿色建筑协会认证专家
2010美国建筑师协会区域和城市规划荣誉奖评委会主席

© Torti Gallas and Partners, Inc. 托尔蒂·格拉斯事务所

As President of Torti Gallas and Partners, Mr. Torti has provided the strong conceptual leadership to bring his firm to national recognition. His firm has been the recipient of 95 national design awards in the last 15 years.

With offices on both coasts and a liaison office in Istanbul, Turkey, he and his partners have built a firm that understands the inextricable tie between urban design and architecture, and between conceptual thinking and creating value for clients and for communities.

Mr. Torti joined the firm in 1973. His conceptual design leadership is key to the success of the firm's projects. As the leader of a market-focused firm, he and his partners have specialized expertise in the development and design of new towns and villages, neighborhoods, homes, main streets, workplaces and civic and institutional buildings.

Prior to joining Torti Gallas and Partners, Mr. Torti was affiliated with NASA and the National Capital Planning Commission, where he worked on numerous designs to rebuild Washington after the 1968 riots. He also was a Principal in an architectural firm in the Midwest and was the director of a non-profit housing and community development corporation.

In recognition of his many design contributions in architecture and urban design, Mr. Torti was elected to the American Institute of Architects College of Fellows in 2001. Mr. Torti is a graduate of the University of Notre Dame with a Bachelor of Architecture degree. He is also a member of the Advisory Council for the School of Architecture at the University of Notre Dame. In 2004, Mr. Torti became a LEED Accredited Professional.

Mr. Torti's teaching credentials include:
-Assistant Professor of Architecture, Catholic University of America, 1970-1973
-Lecturer and Visiting Critic, University of Maryland, University of Virginia, Ohio University, Harvard University

A selected listing of Mr. Torti's recent speaking venues includes:
-American Institute of Architects • The Mayor's Institute on City Design: Northeast
-Urban Land Institute • National Conference of the American Planning
-Congress for the New Urbanism Association
-Multi Housing World Info Expo • Multi-Family Housing Conference
-National Apartment Association • University of Notre Dame
-National Association of Home Builders • University of Maryland International Builders' Show • Andrews University
-The 21st Century Neighborhoods Conference
-University of Miami

作为托尔蒂·格拉斯事务所的主席，托尔蒂先生为公司获得人们的赞誉付出了汗马功劳。他的公司在过去的15年中获得了95项美国国家设计大奖。

公司在美国东西海岸以及土耳其的伊斯坦布尔都有办事处。他和他的合伙人所创立的公司对城市规划与建筑之间的复杂关系、概念思维与为客户和社区创造价值的复杂关系有着独到的见解。

托尔蒂于1973年加入公司。他的概念设计对公司项目的成功至关重要。作为市场化公司的领头人，托尔蒂和他的合伙人专注于开发设计新城镇、社区、住宅、街道、办公区以及市政和学院建筑。

在加入托尔蒂·格拉斯事务所之前，托尔蒂曾经与美国宇航局和首都规划委员会合作。他为1968年暴乱之后华盛顿的重建进行了相当多的设计。他还曾是美国中西部一家建筑公司的总监，并且担任了一个非营利住宅和社区开发公司的总监。

2001年，由于在建筑和城市规划方面做出了杰出贡献，托尔蒂被美国建筑师协会选为学院会员。托尔蒂在美国圣母大学建筑获得了建筑学学士学位，同时也是该大学建筑学院顾问委员会的成员。2004年，他获得了美国绿色建筑协会认证专家资格。

托尔蒂先生的教学背景：
· 1970-1973，美国天主教大学建筑学助理教授
· 美国马里兰大学、弗吉尼亚大学、俄亥俄大学、哈佛大学讲师和客座评论家

托尔蒂先生近期的演讲地点：
· 美国建筑师协会·城市设计市长协会：东北部
· 美国城市土地协会·美国国家规划会议
· 新城市主义协会
· 多户住宅世界信息展览会·多户住宅会议
· 美国国家公寓协会·美国圣母大学
· 美国国家住房建筑商协会·马里兰大学国际建筑商展·安德鲁大学
· 21世纪社区会议
· 迈阿密大学

Alice Tully Hall
爱丽丝杜莉厅

Jury Comments:
This project takes an introverted anti-urban building and engages it with the city, bringing a sense of performance and theater right out to the sidewalk.

评委评语：
项目将低调而内向的建筑与城市结合在一起，将表演和剧院带到了人行道上。

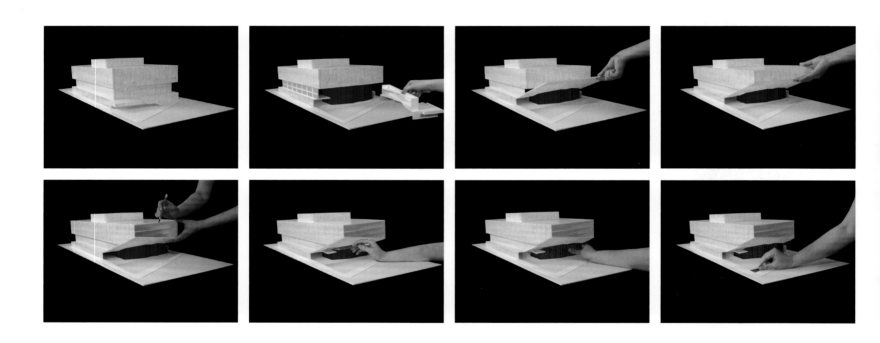

Notes of Interest

The redesign of Alice Tully Hall transforms the venue from a good multi-purpose hall into a premiere chamber music venue with street identity and upgraded functionality for all performance needs. The sloped underside of Juilliard's expansion serves as a canopy framing the hall, its expanded lobby, and box office; the opaque base of Pietro Belluschi's building is stripped away to reveal the hall's outer shell and a shear one-way cable net glass façade puts the hall on display.

Illumination emerges from the wood skin of the hall much the way a bioluminescent marine organism exudes an internal glow. A percentage of wood liner is constructed of translucent custom-molded resin panels surfaced in veneer to match and blend seamlessly with the wood, binding the house and stage with light. Like raising a chandelier signaling the start of the performance, the blush will be part of the choreography.

Consultant: L'Observatoire International, Fisher Dachs Associates
Engineer: ARUP
General Contractor: Turner Construction Co.
Owner: Lincoln Center Development Project

顾问：瞭望国际、费舍尔·达奇斯事务所
工程师：ARUP
总承包商：特纳建筑公司
所有人：林肯中心开发项目

Architect / 建筑师	**Location** / 项目地点	**Photo Credit** / 图片版权
Diller Scofidio + Renfro and FXFOWLE Architects 迪勒·斯科费德+兰弗罗和FX福勒建筑事务所	New York City, New York 纽约州，纽约	© Iwan Baan 伊万·班

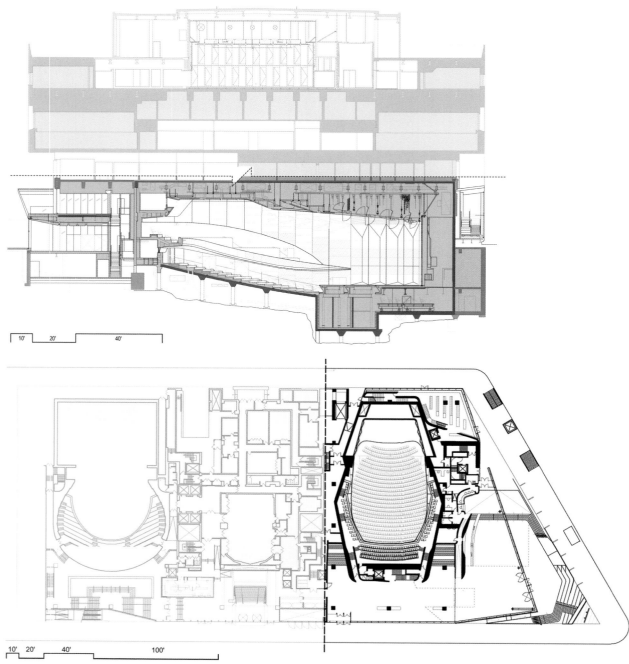

项目特色

爱丽丝杜莉厅的重新设计将其从一个良好的多功能大厅改造成为一个顶级音乐胜地。全新的音乐厅外面光彩夺目，内部设施齐全，能够满足所有演出需求。茱莉亚音乐学院扩建部分的倾斜底座成为了音乐厅、扩展大堂和售票处的华盖；彼得·贝鲁斯奇大楼的不透明底座被剥离，展示了音乐厅的外壳和单向粗网玻璃立面。从音乐厅外壳显露出来的光晕让大厅看起来像一个发光的海洋生物一样。木制衬线的一部分由半透明的塑形树脂板组成，与木材完美地结合在一起，为后台和舞台带来了光亮。就像升起吊灯标志着表演开始一样，光晕也成为了节目编排的一部分。

Austin E. Knowlton School of Architecture
奥斯丁·E·诺尔顿建筑学院

Jury Comments:
This project embodies everything I would want in an architecture building. It is full of unique spaces, an open flexible hall that beckons people to participate, and seems to have surprises around every corner.

评委评语：
项目体现了我所期待的所有建筑学元素。它到处都有独特的空间：一个开放式多功能大厅吸引着人们前来参与其中，每个转角似乎都有惊喜。

Level One, Ground: 一层，地面层：
1. Jury Space and Lecture Rooms — 1. 评审空间和讲座室
2. Café — 2. 咖啡厅
3. Center Space — 3. 中央空间
4. Gallery — 4. 走廊
5. Classroom — 5. 教室
6. Administration — 6. 行政区
7. Front Entry — 7. 正门
8. Forecourt — 8. 前庭
9. Bus Stop/North Gardens — 9. 公交站/北花园
10. South Court — 10. 南院
11. North Porch — 11. 北门廊

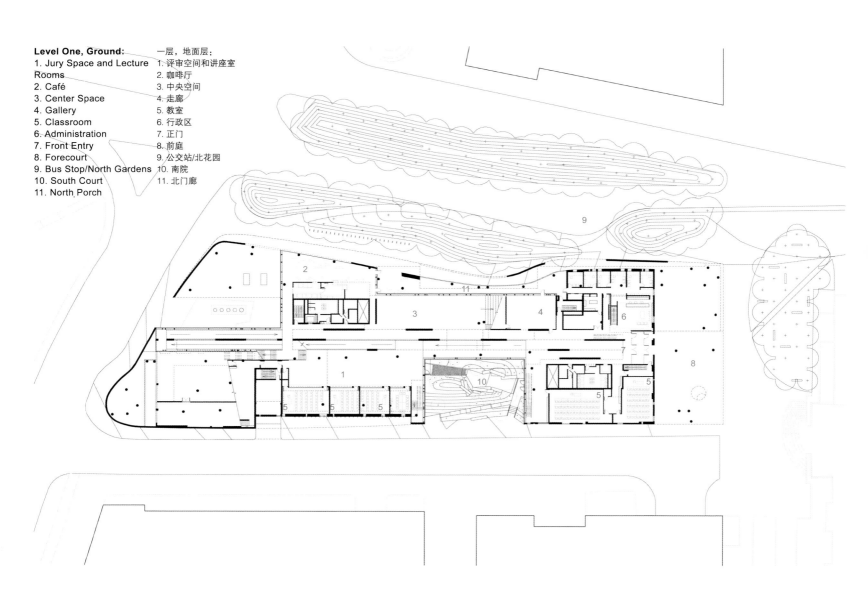

Notes of Interest

The site of the new school of architecture is at the western edge of the old campus, close to the river and the football stadium, at the happily congested corner of West Woodruff Avenue and Tuttle Park Place. Bounded by existing buildings and crossed by major campus pedestrian thoroughfares, the site is a dynamic zone, capable of sustaining a connective architecture and landscape and an inclusive urban form.

Asserting the belief that a school of architecture has a commitment to teach by example to both students within and the community at large, the architectural form and urban positioning of the new school is strategically active and interactive. The building form is generated by enclosing, defining and confronting the spaces and existing buildings of the larger site. Studios overlook the newly captured spaces. Students are in the midst of the urban activity which they will study and will eventually help form and influence.

Consultant: Bird + Bull, Ramon Luminance Design
Engineer: Shelley Metz Baumann Hawk, HAWA Consulting Engineers
General Contractor: P.J. Dick, Inc.
Landscape Architect: HAWA Consulting Engineers
Owner: The Ohio State University

顾问： 小鸟+公牛公司、雷蒙照明设计
工程师： 雪莉·梅兹·鲍曼·霍克、HAWA咨询工程公司
总承包商： P·J·迪克公司
景观建筑师： HAWA咨询工程公司
所有人： 俄亥俄州立大学

Architect /建筑师
Mack Scogin Merrill Elam Architects and WSA Studio
MSME建筑事务所和WSA工作室

Location /项目地点
Columbus, Ohio
俄亥俄州，哥伦布

Photo Credit /图片版权
© Timothy Hursley Photography
狄默思·赫斯利摄影

项目特色

新建的建筑学院位于校园的西部，紧邻河流和足球场，处在西伍德拉夫大道和塔特尔公园的转角处。场地四周环绕着建筑，中间贯穿着校园人行道，是一片活跃的区域，足以容纳一座连接性建筑和景观以及一个内部城市形态。

由于项目要求建筑学院成为建筑学生和整个社区的学习典范，新学院的建筑造型和城市定位均以活跃和互动为基础。建筑造型通过封闭、界定、对比空间和原有建筑结构得来。工作室俯瞰着新建的空间。学生们处于城市活动的中心，他们在那里学习，并将对建筑形成潜移默化的影响。

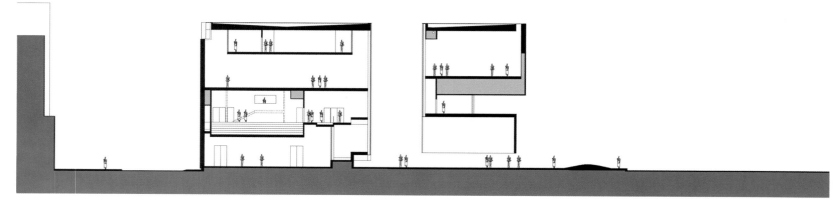

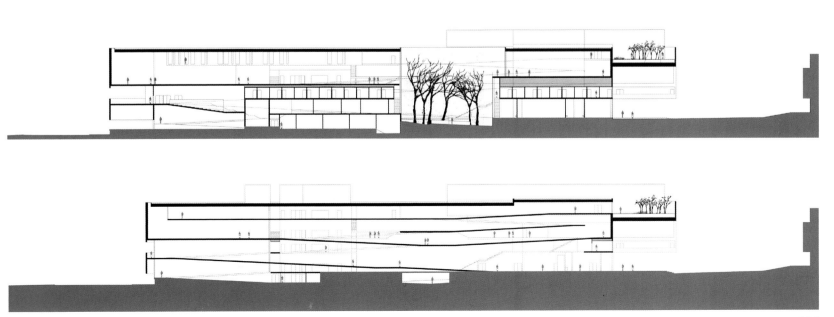
023

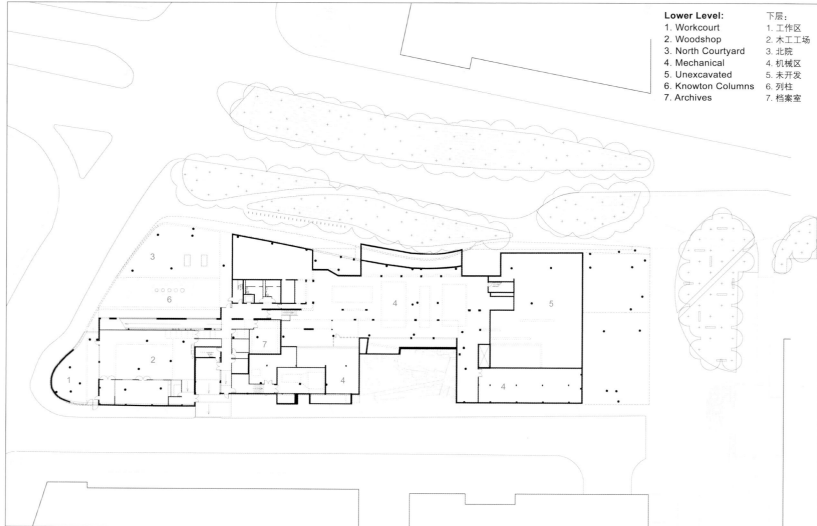

Lower Level:	下层：
1. Workcourt	1. 工作区
2. Woodshop	2. 木工工场
3. North Courtyard	3. 北院
4. Mechanical	4. 机械区
5. Unexcavated	5. 未开发
6. Knowton Columns	6. 列柱
7. Archives	7. 档案室

Beauvoir

美景

Jury Comments:
This is a great example of a successful effort that included finding the funds, doing the research and implementing the work with great skill, discipline and love. As a victim of Hurricane Katrina this project is significant culturally, historically and architecturally. The clarity of the commitment led to a remarkable restoration.

评委评语：
该项目是成功的典范，它成功地找到了资金，进行了研究并以高超的技艺和极大的热情完成了工作。作为卡特里娜飓风的受害者，该项目具有丰富的文化、历史和建筑价值。设计师的翻修工作十分出色。

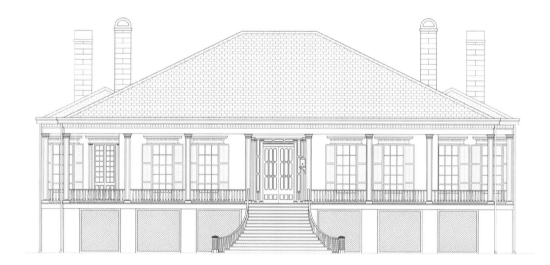

Consultant: George Fore
Engineer: Sparks Engineering, Inc
General Contractor: The Lathan Company Inc.
Owner: Mississippi Division Sons of Confederate Veterans

顾问： 乔治·福尔
工程师： 斯帕克斯工程公司
总承包商： 拉坦公司
所有人： 盟军退伍军人联合会密西西比分部

Notes of Interest

The home that would become known as Beauvoir (meaning "beautiful view") was constructed in 1852 by James Brown, a planter from Madison County, Mississippi. It was owned by the Mississippi Division of the Sons of the Confederate Veterans (SCV) and operated by the State of Mississippi. In 1973, Beauvoir was designated a National Historic Landmark by the National Park Service.

On August 29, 2005, Hurricane Katrina made landfall approximately 60 miles west of Biloxi. The storm surge ripped the piers out from under the porches causing structural failure of the entire front porch and the roof over the front porch, as well as compromising the integrity of the chimneys. Failure of the roof over the front porch also caused extensive damage to adjacent interior ceilings. Ultimately, the mansion had barely survived the worst weather event in its 153 year history with a severely compromised foundation and an overly vulnerable envelope.

项目特色

这座名为"美景"的建筑由詹姆斯·布朗（密西西比州麦迪森郡的一位农场主）建于1852年。建筑为盟军退伍军人联合会密西西比分部所有，由密西西比州政府运营。1973年，美景楼被国家公园管理局认证为国家历史遗址。

2005年8月29日，卡特里娜飓风在距比洛克西以西约60英里处登陆。风暴潮摧毁了门廊下方的支柱，导致整个前廊和前廊屋顶毁坏，同时也破坏了烟囱的完整造型。前廊屋顶的损坏还导致了内部天花板的破损。最终，建筑勉强从自它建成153年以来最大的自然灾害中幸存，仅剩下严重破损的地基和脆弱的外壳。

Architect / 建筑师	**Location** / 项目地点	**Photo Credit** / 图片版权
Albert & Associates Architects 艾伯特建筑事务所	Biloxi, Mississippi 密西西比州，比洛克西	© Sarah A. M. Newton 萨拉·A·M·纽顿

Brochstein Pavilion and Central Quad: Rice University

莱斯大学布洛奇斯坦馆和中央庭院

Jury Comments:
The only non-brick building at Rice University, the Brochstein Pavilion is a deceivingly simple glass, aluminum and steel jewel that solves complex issues on campus and activates the open space of this important circulation area. Its transparency, lightness and immaculate detailing make this structure a refreshing destination on campus.

评委评语：
布洛奇斯坦馆是莱斯大学中唯一一座非砖制建筑，由玻璃、铝材和钢材简单地组成，活跃了校园的重要流通空间。它的通透感、轻盈感和完美的细节设计让自身成为了校园里一个清新自然的场所。

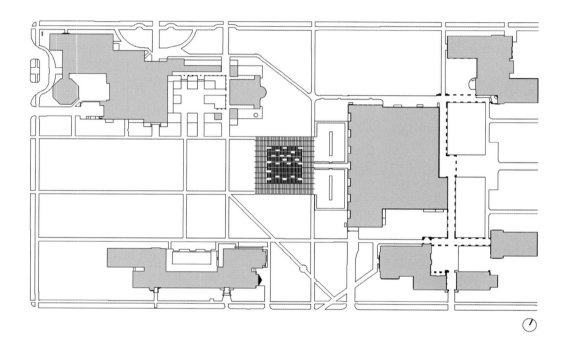

Notes of Interest

Centrally located on Rice University's campus, the Raymond and Susan Brochstein Pavilion was conceived as a destination for students and faculty to interact and share ideas in a relaxed environment. Carefully sited at an important intersection of campus pathways to create a new hub of activity, the Pavilion encourages interaction without interrupting pedestrian movement through campus. The sensitive addition of trees, fountains, and garden seating areas seamlessly blends the new pavilion into the existing quadrangle.

The Brochstein Pavilion is capped by a steel trellis structure which protects the building and extends in all directions to cover and shade the surrounding seating terrace. Shading the entire structure and consisting of an array of small aluminum tubes, the trellis cuts the direct sun by an average of 70 percent. This extensive shade protection reduces the required mechanical cooling load by 30 percent and allows the structure to be open and naturally ventilated throughout much of the year.

Consultant: Fisher Marantz Stone (Lighting), Walter P. Moore (Structural)
Engineer: AltieriSeborWieber LLC, Haynes Whaley Associates
General Contractor: Linebeck Group, LLC
Landscape Architect: The Office of James Burnett
Owner: Rice University

顾问：费舍尔·马兰兹·斯通（灯光）、沃尔特·P·摩尔（结构）
工程师：ASW工程公司、海恩斯·惠利事务所
总承包商：莱恩贝克集团
景观建筑师：詹姆斯·箔内特工作室
所有人：莱斯大学

Architect / 建筑师
Thomas Phifer and Partners
托马斯·菲佛事务所

Location / 项目地点
Houston, Texas
德克萨斯州，休斯顿

Photo Credit / 图片版权
© Scott Frances
斯科特·弗朗西斯

项目特色

雷蒙德和苏珊·布洛奇斯坦馆位于莱斯大学校园的正中，是学生和教职工休闲互动的好去处。项目被精心设置在校园小路的交叉口，打造了一个全新的活动中心，既鼓励人们进行互动，又不会影响行人活动。树木、喷泉和花园让项目锦上添花，与原有的四方院完美地契合在一起。

布洛奇斯坦馆上方的钢铁框架结构保护了建筑，并向四方延伸，为四周的休息平台提供了遮蔽。屋顶结构由一系列小型铝管组成，削减了70%的太阳直射。广泛的遮阳保护减少了30%的机械制冷负荷，让项目在一年中大部分时间都采用开放式自然通风。

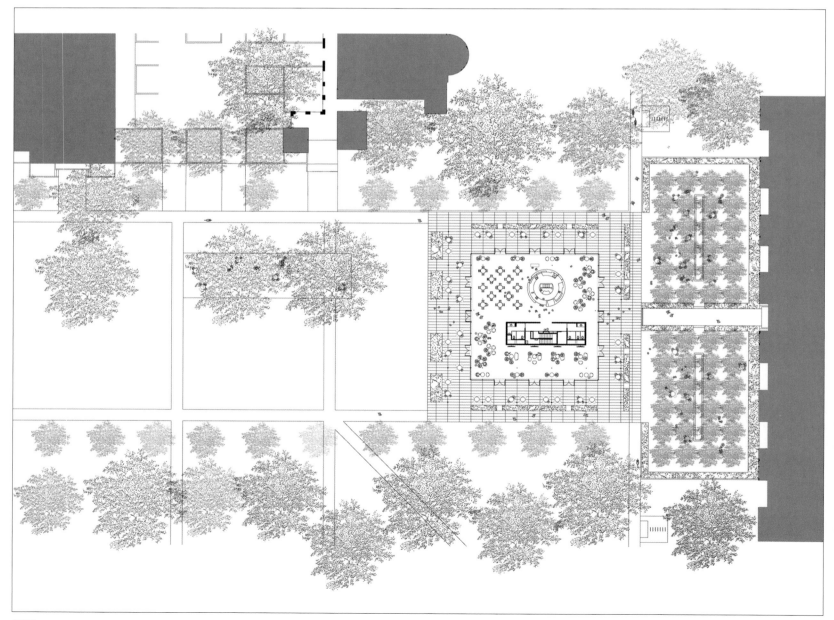

Camino Nuevo High School
卡米诺·尼沃高中

Jury Comments:
Architecturally responsive to program and difficult (thin) site on very tight budget, while solving social and sustainability desires with single loaded exterior circulation and maximum daylighting opportunities... creates an educational haven within a busy urban environment.

评委评语：
项目在狭窄的场地上凭借极少的预算实现了功能需求，同时通过集中荷载外部流通路径和最大化日照量解决了社会和可持续需求，在繁忙的城市环境中打造了一个教育天堂。

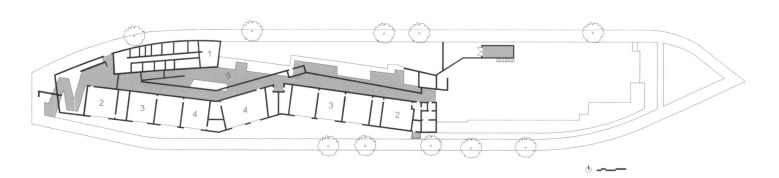

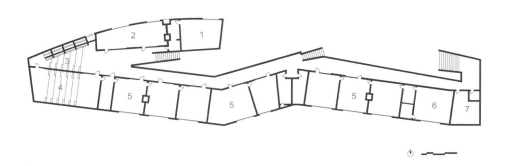

First Floor Plan (Above): 一层平面图（上）：
1. Administration   1. 行政区
2. Science Lab   2. 科学实验室
3. Classroom   3. 教室
4. Art Classroom   4. 艺术教室
5. Court yard   5. 庭院

Second Floor Plan (Left): 二层平面图（左）：
1. Meida Center   1. 媒体中心
2. Science Lab   2. 科学实验室
3. Outdoor Amphitheater 3. 露天剧场
4. Auditorium   4. 礼堂
5. Classroom   5. 教室
6. Science Lab   6. 科学实验室
7. Storage   7. 储藏室

Notes of Interest

This charter high school houses 500 students in Silver Lake, a multi-cultural community adjacent to downtown Los Angeles. It is the third project, in a series of four that the firm has designed for the charter school client. The schools were launched by a nonprofit community development corporation to provide small, focused schools for children in a dense and underserved urban Los Angeles neighborhood. In 2000 and 2003 the office completed an elementary and a middle school on a single block campus in MacArthur Park.

A winding form for the 30,000 square foot, 18-classroom building maximizes the space available on the oddly shaped site. Additionally, by single-loading the one, main classroom building, two important social and sustainable functions were accomplished with simple solutions: direct visual connections are established between the classrooms and the inner courtyard and natural light flows into each classroom from both the windows on the street side and courtyard side. This courtyard has become the hub of the school.

Consultant: Konsortum 1, Pfeiler and Associates
Engineer: John Labib + Associates, Tsuchiyama Kaino Sun & Carter
General Contractor: Turner Special Projects
Landscape Architect: Ah be
Owner: Pueblo Nuevo Development

顾问： 肯索尔特穆1、皮菲勒事务所
工程师： 约翰·拉比伯事务所、土山新·孙和卡特工程公司
总承包商： 特纳特殊项目公司
景观建筑师： 阿尔比
所有人： 普韦布洛·尼沃开发公司

Architect / 建筑师
Daly Genik
达利·吉尼克

Location / 项目地点
Los Angeles, California
加利福尼亚州，洛杉矶

Photo Credit / 图片版权
© Tim Griffith
蒂姆·格里菲斯

项目特色

这家位于银湖区（紧邻洛杉矶市中心的一个多文化融合的社区）的特许公立高中拥有500名在校学生。它是设计公司所设计的第三个特许公立学校项目。这些学校由一家非营利社区发展公司投资，专门在密集的洛杉矶城市区域提供小型学校。在2000年和2003年，该公司在麦克阿瑟公园的一个街区已经建成了一所小学和一所初中。

弯曲造型的教学楼总面积2,787平方米，共有18间教室，古怪的造型将内部空间最大化。此外，集中荷载让主教学楼通过简单的解决方案实现了社会和可持续价值：设计师在教室和内部庭院之间建立了直接视觉联系，并且让自然光通过街道和庭院两侧的窗口射入每间教室。这个庭院还成为了校园的中心。

Campus Restaurant and Event Space
园区餐厅和活动空间

Jury Comments:
Crisp, elegant and ordered with a fantastic floating canopy that engages the campus landscape, this project brings together company staff of all types into a light-filled, open gathering place.

评委评语：
简洁、优雅而有序的建筑与奇妙的悬浮穹顶为园区景观增添了活力，项目让公司员工聚集在这个光亮而开放的集会空间。

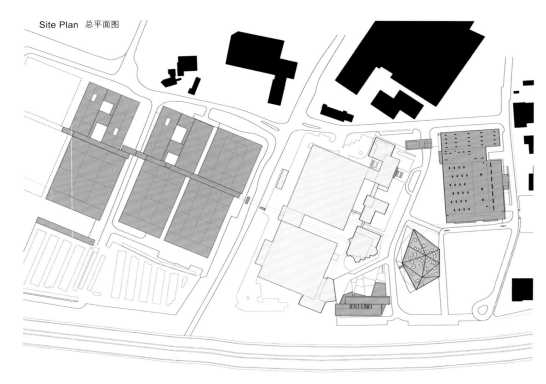

Site Plan 总平面图

Notes of Interest
This pavilion provides a new central cafeteria and event space for a Stuttgart-based industrial campus. It is the new social center for the company for both blue- and white-collar workers. The pavilion, with seating for 700, enables the company's 2,000 employees to lunch around three time slots in the large central space with reserve seating in a mezzanine. When programmed for events the space functions as an auditorium with seating for 800.

A floating roof hovers over the central dining space and mezzanine that are placed in an excavated hollow. The intention was to create a polygonal leaf-like canopy that wide-spans over column-groups. Aesthetically, the roof as a fifth facade, is carefully organized with skylights, and air-vents, as it is highly visible from the mid-rise office buildings adjacent to it.

Neither a factory nor an office building, this freestanding pavilion introduces a new typology to the campus. Urbanistically the new restaurant helps to complete spatially the entrance courtyard. Formally its crystalline pentagon plan is a continuation of the crystalline ground plans of the new office building to which it is adjacent.

Consultant: Transsolar Energietechnik GmbH (Climate), Gassmann + Grossmann (Management)
Engineer: Werner Sobek Stuttgart (Structural; Facade) Schuckertstrasse 27
Landscape Architect: Buero Kiefer
Owner: TRUMPF GmbH + Co.KG

顾问：超日工程公司（气候）、加斯曼+格罗斯曼（管理）
工程师：维尔纳·索贝克·斯图加特（结构；外立面）、舒克尔特斯27号
景观建筑师：布埃罗·基弗
所有人：通快公司

项目特色
该项目位于斯图加特的工业园区内，提供了一个全新的咖啡和活动空间。它是公司为蓝领和白领所提供的全新的社交中心。建筑可容纳700人，足以让公司的2,000名员工分三批进行午餐。在举办活动时，整个空间可被作为礼堂，容纳800人。

悬浮在中心餐厅上方的屋顶位于中空结构之上，打造了一个横跨列柱的多边形树叶状穹顶。从美学观点看，屋顶是第五个外立面，上方的天窗和气孔在旁边的中层建筑上清晰可见。

这座独立建筑既非工厂也非办公楼，为校园区引入了全新的建筑类型。在规划上，餐厅帮助完善了入口庭院的空间布局。在造型上，它的五角形透明规划是其紧邻的办公楼的延续。

Architect / 建筑师
Barkow Leibinger Architects
巴考·利宾格建筑事务所

Location / 项目地点
Stuttgart, Germany
德国，斯图加特

Photo Credit / 图片版权
© Amy Barkow
艾米·巴尔考

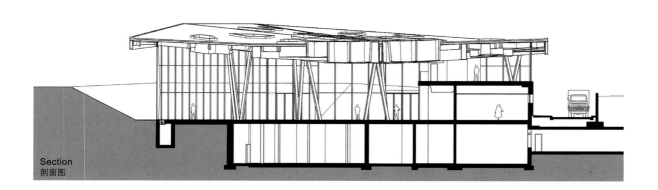

Section
剖面图

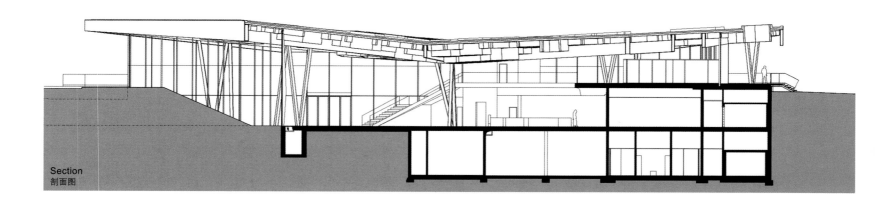

Section
剖面图

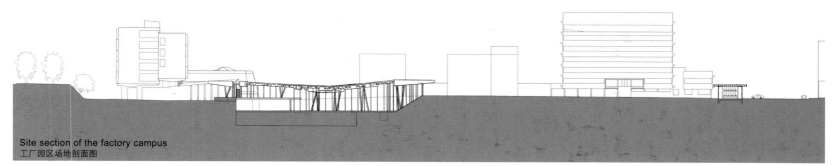

Site section of the factory campus
工厂园区场地剖面图

CAMPUS RESTAURANT AND EVENT SPACE
园区餐厅和活动空间

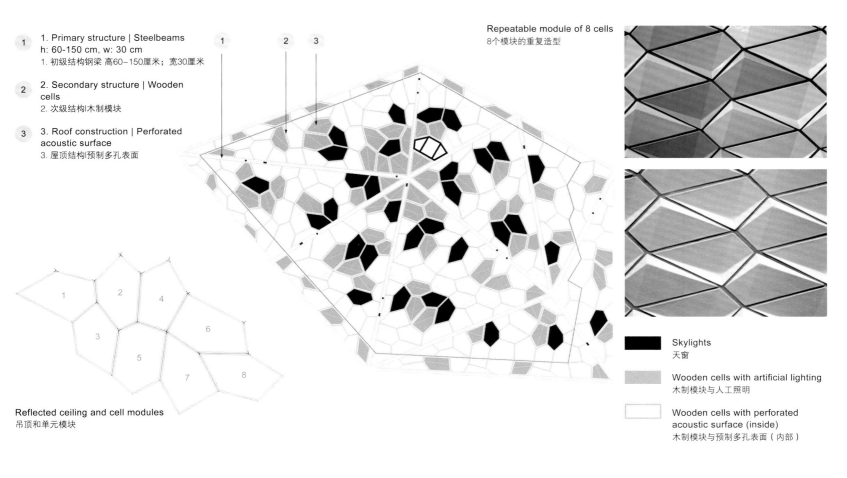

1. Primary structure | Steelbeams
   h: 60-150 cm, w: 30 cm
   初级结构钢梁 高60–150厘米；宽30厘米
2. Secondary structure | Wooden cells
   次级结构|木制模块
3. Roof construction | Perforated acoustic surface
   屋顶结构预制多孔表面

Repeatable module of 8 cells
8个模块的重复造型

Reflected ceiling and cell modules
吊顶和单元模块

■ Skylights
天窗

▨ Wooden cells with artificial lighting
木制模块与人工照明

□ Wooden cells with perforated acoustic surface (inside)
木制模块与预制多孔表面（内部）

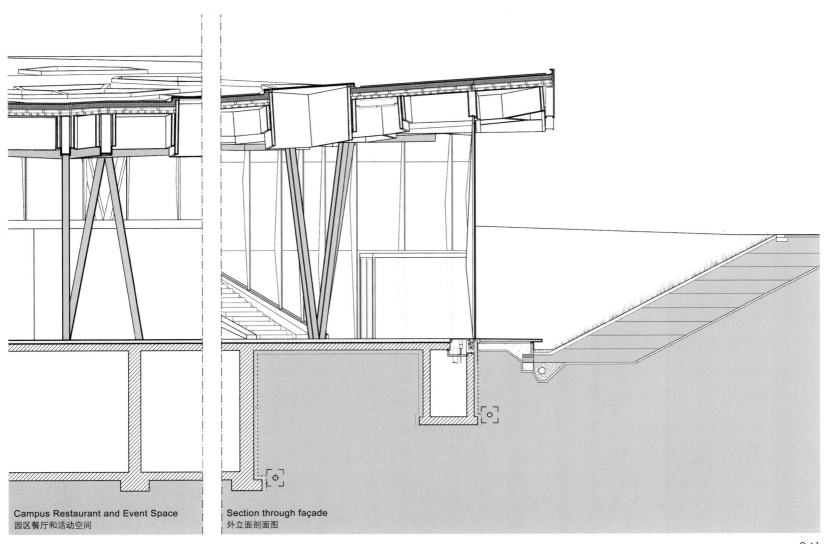

Campus Restaurant and Event Space
园区餐厅和活动空间

Section through façade
外立面剖面图

041

Floor Plan: Mezzanine (+ 1m):
1. Restaurant
2. Terrace
3. Caféteria
4. Air supply
5. Leaving air, return cooling tower

平面图：中层楼（+1米）：
1. 餐厅
2. 平台
3. 自助餐厅
4. 空气供给
5. 出气孔，回到冷却塔

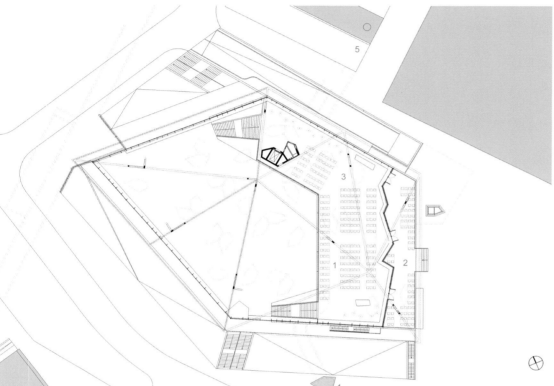

Floor Plan: Level 1 (- 4m):
1. Delivery
2. Storage
3. Kitchen
4. Foodcounters
5. Restaurant and Auditorium
6. Lounge
7. Tunnel connection of factory campus

平面图：一层（-4米）：
1. 配送间
2. 储藏室
3. 厨房
4. 食品柜台
5. 餐厅和礼堂
6. 休息室
7. 与工厂园区的隧道连接

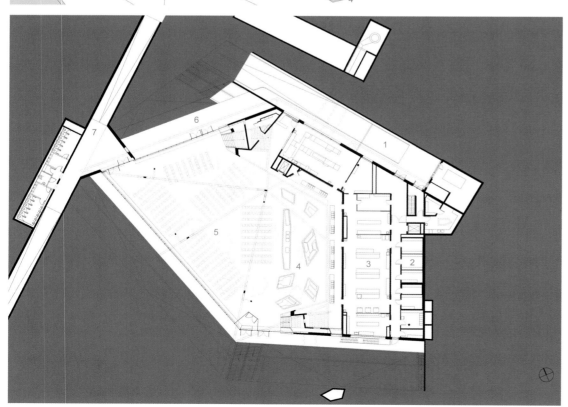

Macallen Building
麦卡伦大厦

Jury Comments:
This is a bold architectural statement in which the architects and the client did not shy away from taking risks. The building is inventive and at times, ingenious.

评委评语：
建筑师和委托人不怕承担风险，进行了大胆的建筑表述，建筑创新而独树一帜。

Notes of Interest
As a pivotal building in the urban revitalization of South Boston, the Macallen's design required a reassessment of conventional residential typologies to produce an innovative and sustainable building that worked within a developer's competitive budget. Occupying a transitional site that mediates between highway off-ramps, an old residential fabric, and an industrial zone, the building negotiates different scales and urban configurations through varied spatial conditions, various ways of reacting to the public sphere, and different material and façade articulations.

On the western end, the building responds to the highway and Boston skyline with a glass curtain wall yielding panoramic views for the residents inside. On the eastern end, brickwork mirrors that of the neighborhood's building fabric, extending the logic of the storefront and pedestrian scale elements. On the north and south façade, bronzed aluminum panels reflect the industrial zone and express the structural system within.

The Macallen is fully integrated – in structure, and sustainability – and is replete with sustainable features to make it the first LEED gold certified building of its type in Boston.

General Contractor: Bovis Lend Lease
Landscape Architect: Landworks Studio
Owner: Pappas Enterprises, Inc.
总承包商：宝维士联盛公司
景观建筑师：地景工作室
所有人：帕帕斯公司

Architect /建筑师
Office dA, Inc. and Burt Hill
dA工作室和伯特·希尔

Location /项目地点
Boston, Massachusetts
马萨诸塞州,波士顿

Photo Credit /图片版权
© John Horner Photography
约翰·霍纳摄影

项目特色

作为南波士顿城市振兴的一座关键建筑，麦卡伦大厦的设计重新评估了传统住宅类型，打造了一座创新的可持续建筑。大厦占据了公路坡道、旧住宅区和工业区的过渡区域，通过各种各样的空间条件调节了不同规模和城市配置。它利用各种方式来与公共区域进行互动。

建筑的西端通过玻璃幕墙与公路和波士顿的天际线相互联系，为内部的居民提供了优美的景致。建筑西端的砖砌结构反映了周边建筑的外观，延伸了店面和人行道的空间。南北两侧外立面的镀铜铝板反映了工业区的特征，也展现了其内部结构。

麦卡伦大厦结合了建筑结构和可持续性，是波士顿第一座获得绿色建筑金奖认证的同类建筑。

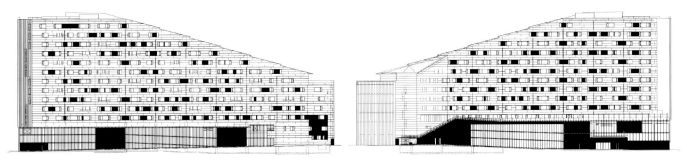

Outpost
前哨

Jury Comments:
A severely simple building evokes ages-old images of oases and paradisiacal gardens transported to the American west. Industrial materials, while rough, are thoughtfully detailed and crafted, and are handled with great skill to make a sublime place.

评委评语:
极简的建筑唤醒了人们对古老的绿洲的回忆,天堂花园则带领人们到达了美国西部。粗犷的工业化材料经过了精心设计和制作,运用高超的技巧打造了美好的空间。

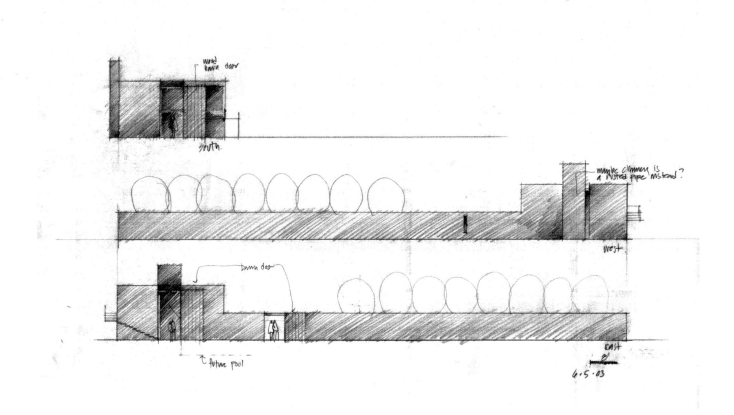

Notes of Interest
Set in the remote and harsh high desert landscape of Idaho, Outpost is an artist live/work studio and sculpture garden for making and displaying art. An important aspect of the complex is the protected "paradise garden", which is separated from the wild landscape by thick masonry walls. The materials used in the structure, including concrete block, car-decking, and plywood, require little or no maintenance, and are capable of withstanding the extreme weather that characterize the desert's four seasons.

Outpost's compactness limits site impact and reinforces the desire to be outside. The architects chose a readily available construction material – concrete block – for the primary structure; commercial builders were able to quickly and cheaply assemble the building. Interiors are exposed and unfinished. In a windy environment, the enclosed garden provides protection to develop a cultured space. Nothing outside the walls is modified. The footprint of the building is the limit of intrusion into the landscape – a simple, clearly defined space within the landscape.

Engineer: Monte Clark
General Contractor: Upham Construction
Owner: Jan Cox
工程师: 蒙特·克拉克
总承包商: 阿伯汉姆建筑公司
所有人: 简·考克斯

Architect / 建筑师	Location / 项目地点	Photo Credit / 图片版权
Olson Kundig Architects 奥尔森·昆丁建筑事务所	Central Idaho 中爱达荷州	© Tim Bies 蒂姆·比亚斯

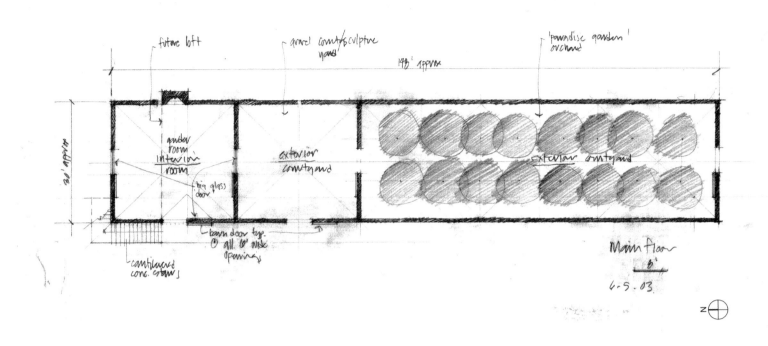

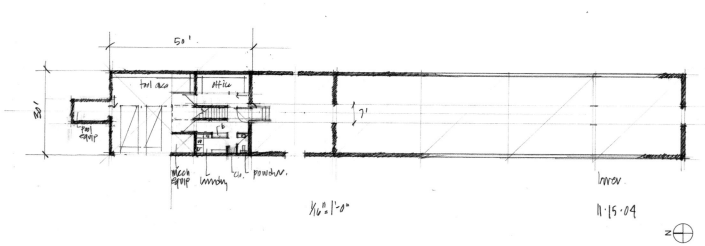

项目特色

前哨坐落在爱荷华州一片荒凉的沙漠景观之中,包括专门供艺术家生活、工作的工作室和用于制作和展示艺术品的雕塑花园。项目的重要组成部分之一是"天堂花园",它通过厚重的石墙与荒野景观隔开。建筑结构中运用的材料包括混凝土砌块、汽车甲板和胶合板。它们基本无需维护,同时能抵挡沙漠四季的极端气候。
前哨紧凑的结构限制了场地对环境的冲击并增强了对外部空间的渴望。建筑师选择了随时可用的建筑材料——混凝土砌块来建造主体结构。施工团队可以快速并高效地组装建筑。室内空间裸露在外,未经加工。在大风环境中,封闭的花园为开发精致空间提供了保护。墙壁外的空间都未经修饰。建筑对景观的空间占用极小——是景观环境中一个简单、明晰的空间。

Serta International Center
舒达国际中心

Jury Comments:
A corporate headquarters that goes beyond trying to project a "powerful" corporate image... Not only is the building beautiful and functional, but it is a testament to Serta's commitment to their employees and providing them with an inspiring workspace and to the preservation of the environment around them.

评委评语:
这是一个远远超越了打造"强大"企业形象的公司总部……建筑美观而兼具功能价值,并且展现了舒达公司对员工的承诺,为他们提供了一个鼓舞人心的办公空间,还保护了周边的自然环境。

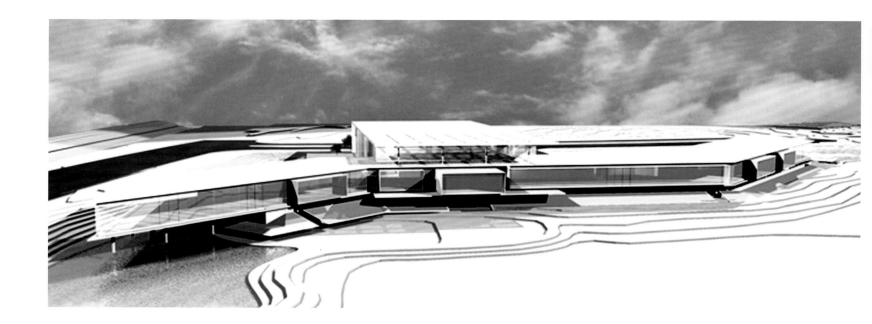

Notes of Interest

The project is a 90,000-square-foot world corporate headquarters for Serta International Mattress Company located on a 20-acre Illinois prairie site. The program combines a 65,000-square-foot office wing and a 25,000-square-foot-high bay R&D facility.

The building (700 feet long x 67 feet wide) has been designed to facilitate equal access to natural light, ventilation, and views of the wetlands for all employees. To accomplish this, the floor plan has been layered from a glass edged public circulation path on the east, to an open office area on the west toward the wetlands. The design also takes advantage of the variations in the topography of the site, to weave together the building and the landscape into a strong holistic composition. The design intent of the building is to float lightly on the landscape, reinforcing the notion of environmental sustainability and echoing the lines of the prairie.

Engineer: Epstein
General Contractor: G.A. Johnson and Sons
Landscape Architect: Jacob/Ryan Associates
Owner: Serta International
工程师:爱普斯坦
总承包商:G·A·约翰逊集团
景观建筑师:雅各布/赖安事务所
所有人:舒达国际

Architect / 建筑师
Epstein | Metter Studios
爱普斯坦|米特尔工作室

Location / 项目地点
Hoffman Estates, Illinois
伊利诺伊州,霍夫曼特斯

Photo Credit / 图片版权
© Metter Photography, © Epstein, © Andrew Metter
米特尔摄影、爱普斯坦、安德鲁·米特尔

项目特色

作为舒达国际马特里斯公司的总部,项目总面积8,361平方米,坐落在伊利诺伊州一块20英亩的草原上。项目包含一座6,039平方米的办公楼和一个2,322平方米的高顶棚研发中心。

建筑长213米,宽20米,力求让所有员工都能享受同等的自然采光、通风以及湿地的开阔视野。为了实现这一目标,建筑的平面规划从东侧的玻璃边公共走道开始分层,直达西侧朝向湿地的开放式办公区。设计还充分利用了场地的地形,让建筑与景观交织在一起。建筑的设计意图是使其轻盈地飘浮于景观之上,增强环境可持续性并且与草原的线条遥相呼应。

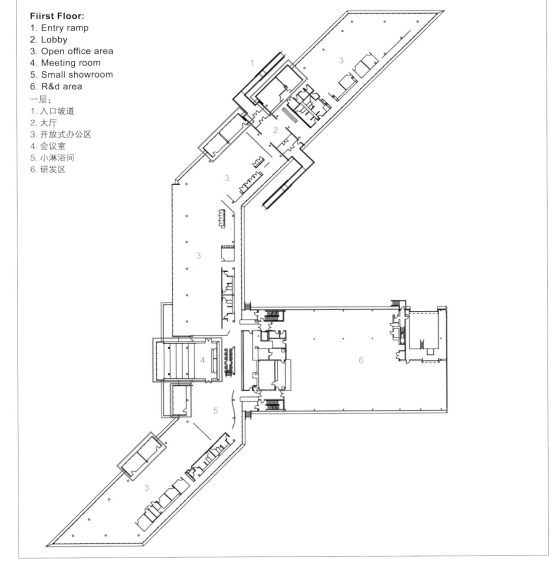

First Floor:
1. Entry ramp
2. Lobby
3. Open office area
4. Meeting room
5. Small showroom
6. R&d area

一层:
1. 入口坡道
2. 大厅
3. 开放式办公区
4. 会议室
5. 小淋浴间
6. 研发区

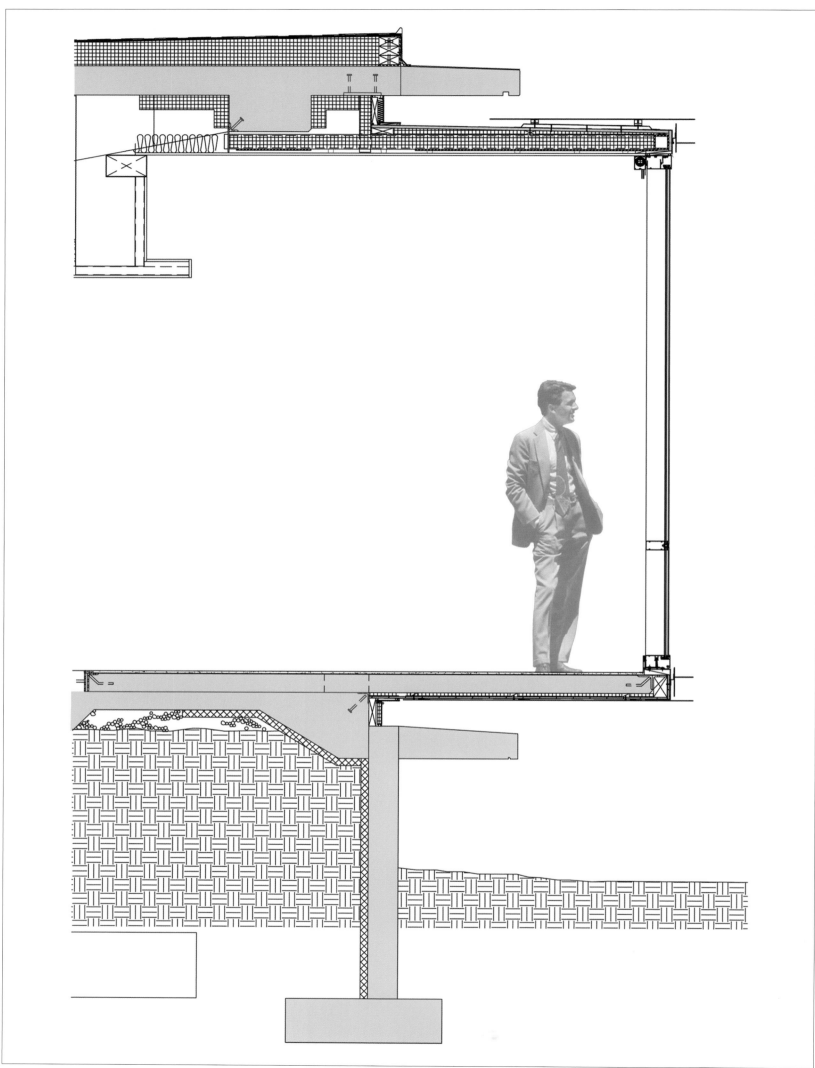

Skirkanich Hall
斯科尔卡尼奇厅

Jury Comments:
There is a thoughtful use of materials, genius in vertical circulation, solid programmatic resolution... both delicate and dramatic, all in all a beautiful project.

评委评语：
设计师在材料、垂直交通和功能方案方面精心规划，打造了精致而引人注目的美观项目。

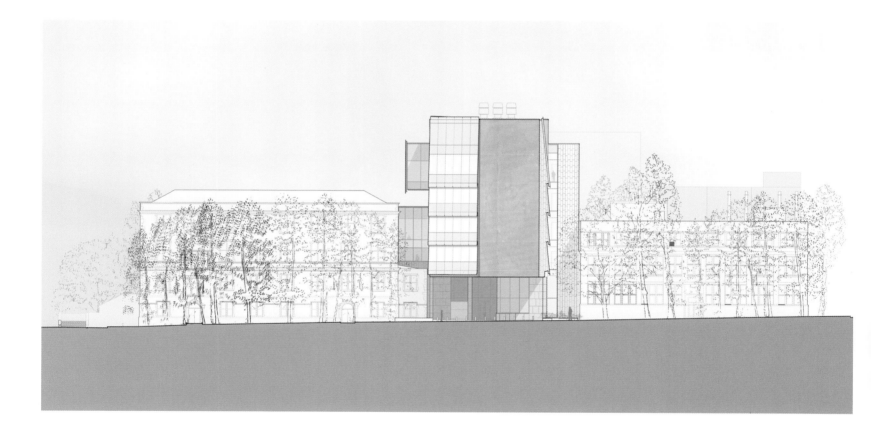

Notes of Interest

Located in the heart of the University of Pennsylvania, Skirkanich Hall is home to the Bioengineering Department. It is an infill building that functions as a connector by creating a new public quadrangle and entry for the School of Engineering and Applied Sciences. Movement and interaction is emphasized with generous circulation spaces that offer places to sit and gather.

The building is cantilevered over the street and descends twenty feet below grade to minimize vertical impact. An open atrium continues up from the ground through five floors above. The laboratories are placed on either side. A vibrant yellow tile with a changing pattern enhances the core space.

To subtly stand out from the red brick buildings next door, a new kind of brick was developed through an extensive process to balance texture, color, durability, and stability. The mossy green colored brick changes with the light of the day. Giant glass shingles contrast the density of the surrounding masonry and bring filtered light into the laboratories.

Associate Firm: The Rose + Guggenheimer Studio
Consultant: GPR Planners Collaborative, Fisher Marantz Stone
Engineer: Ambrosino DePinto & Schmeider, Severud Associates
General Contractor: Skanska USA Building, Inc.
Landscape Architect: Edmund Hollander
Owner: The University of Pennsylvania

合作公司： 罗斯+古根海姆工作室
顾问： GPR规划公司、费舍尔·马兰士·斯通
工程师： 阿姆布罗西诺·品托&施米德尔、瑟维拉德事务所
总承包商： 斯堪斯卡美国建筑公司
景观建筑师： 艾德蒙·霍兰德尔
所有人： 宾夕法尼亚大学

Architect / 建筑师
Tod Williams Billie Tsien Architects
陶德·威廉姆斯·比利·西恩建筑事务所

Location / 项目地点
Philadelphia, Pennsylvania
宾夕法尼亚州，费城

Photo Credit / 图片版权
© Michael Moran
迈克尔·墨兰

项目特色

斯科尔卡尼奇厅位于宾夕法尼亚大学的中心，是生物工程部门的所在地。这座建筑为工程和应用科学学院打造了一个全新的公共庭院和入口。宽敞的流通空间提供了休息和集会场所，突出了建筑的运动和互动功能。建筑呈悬臂结构悬于街面之上，同时也采用了地下结构，将垂直空间的环境影响缩到最小。开放式中庭从地面向上延伸了五层楼。实验室分设中庭两侧。充满活力的黄色地砖和不断变换的图案提升了核心空间效果。为了与旁边的红砖建筑相区别，设计师通过对材质、色彩、耐用性和稳定性的研究研发出一种新型砖。苔绿色砖块随日光的变换而变换。巨大的玻璃瓦与周边的石砌建筑形成鲜明对比，为实验室带来充足的光线。

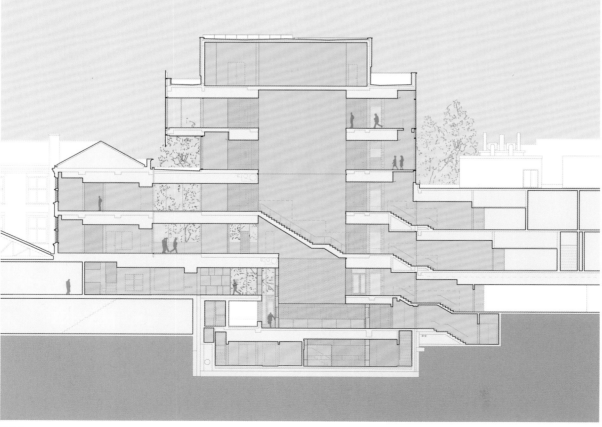

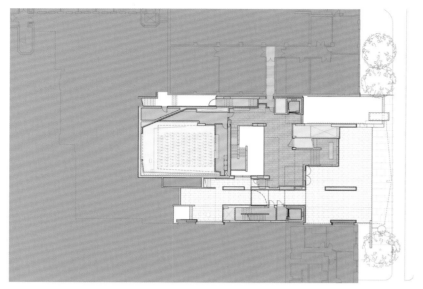
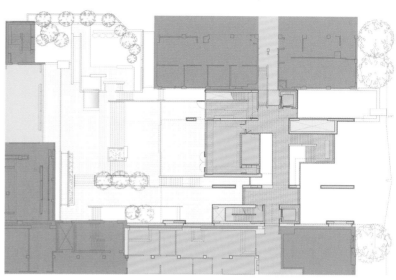
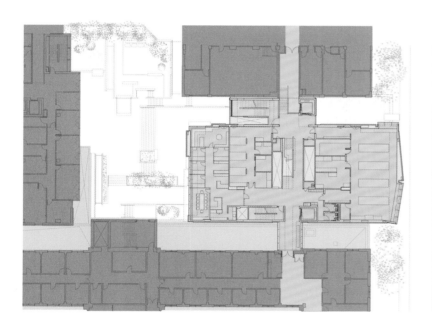
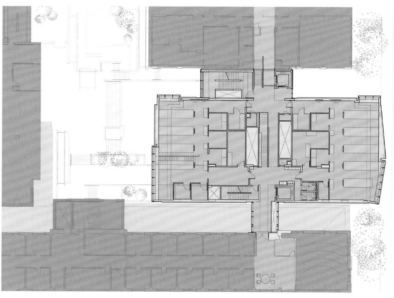
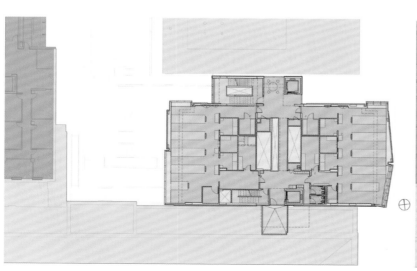
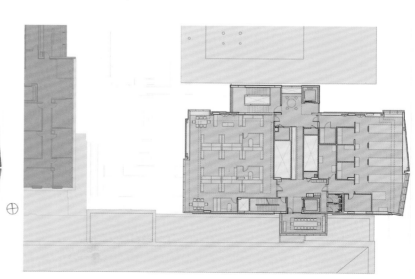

066

Step Up on 5th
登上第五层

Jury Comments:
Step Up on 5th is a very conscious piece of architecture in general. Not only a refuge for the homeless and mentally disabled, it is also an incredibly dense and sustainable piece of architecture. It's an example of a project with a lot of thought and care put into the design of it for the betterment of the community and the inhabitants.

评委评语:
登上第五层是一座非常清晰明确的建筑。它不仅是患有精神障碍的无家可归人士的住处,还是一座无以伦比的密集型可持续建筑。它是精心设计的典范,改善了社区和住户的生活条件。

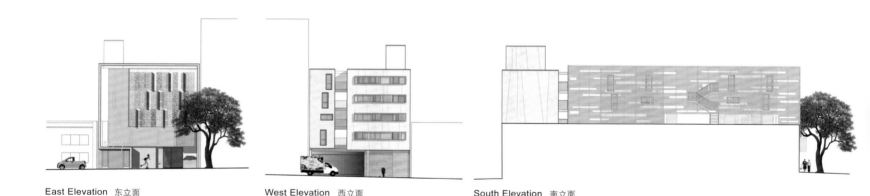

East Elevation 东立面　　West Elevation 西立面　　South Elevation 南立面

Notes of Interest

The new 46-unit mixed-use building provides a home and support services for the mentally disabled homeless population.

The main façade made from custom water jet anodized aluminum panels creates a screen that sparkles in the sun and glows at night, while also acting as sun protection and privacy screens. The material reappears as a strategic arrangement of screens lending a subtle rhythm to the exterior circulation. South-facing walls filter direct sunlight creating a sense of security for the emotionally sensitive occupants.

The project incorporates energy efficient measures that exceed standard practice, optimize building performance, and ensure reduced energy. The design emerged from close consideration and employment of passive solar design strategies that make this building 50 percent more efficient than a conventionally designed structure. While California has the most stringent energy-efficient requirements in the United States, the building exceeds LEED standards and state mandated Title 24 energy measures by more than 30 percent.

Consultant: Laschober + Sovich, Helios International, Inc.
Engineer: John Martin and Associates, IBE
General Contractor: Ruiz Brothers
Landscape Architect: LAND Studio
Owner: Step Up on Second

顾问:拉舍波尔+索维奇、赫利厄斯国际公司
工程师:约翰·马丁事务所、IBE
总承包商:鲁伊兹兄弟
景观建筑师:LAND工作室
所有人:登上第二层

Architect / 建筑师	**Location** / 项目地点	**Photo Credit** / 图片版权
Pugh + Scarpa 佩什+斯卡帕	Santa Monica, California 加利福尼亚州，圣塔莫尼卡	© John Linden 约翰·林登

项目特色

这座46户商住两用建筑为患有精神障碍的无家可归人士提供住处和相关服务。
建筑的主外立面由定制阳极电镀铝板制成，打造了在既能在白天闪光、又能在夜晚发光的外屏。外立面还能起到遮阳和保护隐私的作用。外屏材料的策略布置为建筑外部带来微妙的韵律感。朝南的墙面能过渡直接阳光，为情绪敏感的住户提供了安全感。
项目的节能策略超越了常规标准、优化了建筑性能并且大大减少了能耗。设计所采用的被动式太阳能设计让建筑比传统结构节能50%。加州拥有美国最严格的节能标准，但是建筑已经超过了绿色建筑标准和加利福尼亚州24号能源措施所要求的30%。

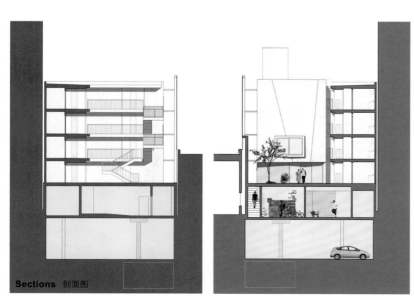

Sections 剖面图

Ground Level Floor Plan (Below):
1. Residential entry lobby
2. Mail boxes
3. Public restroom
4. Elevator
5. Cookline
6. Pot wash
7. Cold prep
8. Refridgerator
9. Dry storage
10. Trash room
11. Parking

一层平面图（下）：
1. 住宅入口大厅
2. 邮箱
3. 公共洗手间
4. 电梯
5. 流水线
6. 洗涤间
7. 冷餐准备室
8. 冰箱
9. 干仓库
10. 垃圾房
11. 停车场

Fourth Level Floor Plan (Left Buttom):
1. Unit living space
2. Unit kitchen
3. Unit bathroom
4. Community room
5. Laundry room
6. Trash room
7. Elevator
8. Perforated screen
9. Manager's unit

四层平面图（左下）：
1. 单元居住空间
2. 单元厨房
3. 单元浴室
4. 公共室
5. 洗衣房
6. 垃圾房
7. 电梯
8. 镂空屏风
9. 管理室

Second Level Floor Plan (Below):
1. Unit living space
2. Unit kitchen
3. Unit bathroom
4. Community room
5. Laundry room
6. Trash room
7. Elevator
8. Perforated screen
9. Manager's unit

二层平面图（下）：
1. 单元居住空间
2. 单元厨房
3. 单元浴室
4. 公共室
5. 洗衣房
6. 垃圾房
7. 电梯
8. 镂空屏风
9. 管理室

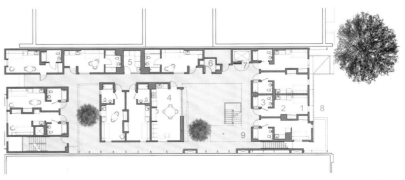

TKTS Booth and the Redevelopment of Father Duffy Square
TKTS 售票亭和杜菲神父广场重建工程

Jury Comments:
With its elegant conception and realization, its refined design stands up to the cacophony of Times Square; this is as much a 21st Century art piece as a building.
The very idea of the building is playful: a structure for selling tickets to shows while also being a vehicle for watching the very "theater" of activity in Times Square. It will be a catalyst for ongoing pedestrian enhancements of the Square.
A simple building, whimsical in nature… it is efficient and functional… sculptural and energetic.

评委评语：
优雅的设计理念和实践让项目在时代广场中脱颖而出，既是一座建筑，又是一件21世纪的艺术品。
建筑设计理念充满趣味：它既是售票处又是观察时代广场大舞台看台。它将成为广场改造步行区的催化剂。
简洁的建筑，异想天开……它高效而实用，精致而富有活力。

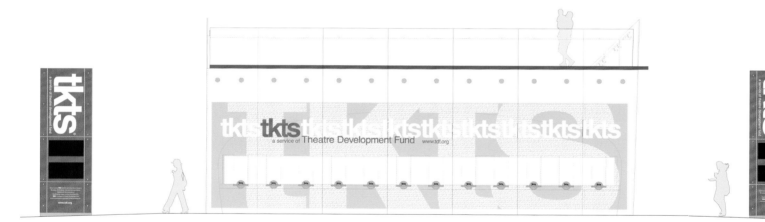

Notes of Interest

The new TKTS Booth, including the redevelopment of Father Duffy Square, creates a new center for Times Square, one of the world's most popular and iconic destinations. The project began in 1999 with a design competition to re-design the popular TKTS booth. While the competition brief simply requested designs for a small scale architectural structure to replace the existing ticket booth, the concept-winning design reframed the problem as one requiring a broader urban design response to invigorate and provide a center for Times Square, and won the competition.

In 2001, the client commissioned a firm to conduct a feasibility study to evaluate the conceptual design scheme. The final design was informed and inspired by the original concept but also used a distinctly 21st Century set of approaches: glass would now be employed as the TKTS Booth's sole structural component for the steps and the TKTS Booth itself would be free-standing within the glass enclosure. The transformation of the public space of Father Duffy Square by the Plaza architect allows for increased pedestrian traffic and more prominence for Father Duffy's commanding statue.

Consultant: Dewhurst MacFarlane and Partners, Fisher Marantz Stone, Bresnan Architects, PC
Engineer: Dewhurst MacFarlane and Partners, Schaefer Lewis Engineers, PC, DMJM Harris, Haran Glass, with IG Innovation Glass LLP
Landscape Architect: Judith Heintz Landscape Architects
Owner: Times Square Alliance, Theatre Development Fund, Coalition for Father Duffy and City of New York

顾问：杜赫斯特·麦克法兰事务所、费舍尔·马兰士·斯通、布雷斯南建筑事务所
工程师：杜赫斯特·麦克法兰事务所、施明贤工程公司、DMJM·哈里斯、哈兰玻璃与IG创意玻璃公司
景观建筑师：朱迪丝·亨斯景观设计事务所
所有人：时代广场联盟、剧院开放基金、杜菲神父和纽约市联合会

Architect /建筑师	**Location** /项目地点	**Photo Credit** /图片版权
Perkins Eastman, Choi Ropiha, and PKSB Architects	New York City	© Paúl Rivera
珀金斯·伊斯曼、蔡·罗比哈、PKSB建筑事务所	纽约州,纽约	保罗·里维拉

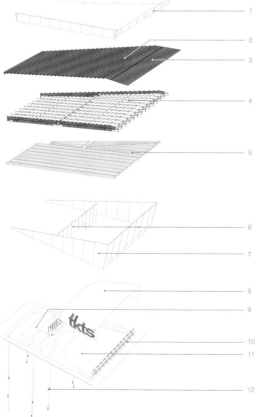

项目特色

TKTS售票亭与杜菲神父广场重建工程共同为时代广场——世界上最受欢迎的标志景点之一——打造了全新的中心。项目始于1999年的一次TKTS售票亭设计竞赛。竞赛要求重新设计伊尔小规模建筑结构来替代原有的售票亭,但是获奖设计重新构造了项目,打造了更广泛的城市设计来活跃时代广场。

2001年,委托人委托一家公司对这一设计方案进行可行性研究。最终的设计从原始方案中获得了灵感,同时兼具21世纪的新元素:TKTS售票亭的底部结构采用了玻璃台阶而自身也被密封在独立的玻璃外壳之中。杜菲神父广场公共区域的改造增加了步行空间,进一步突出了杜菲神父雕像的形象。

1. Glass Balustrades
2. Laminated Glass Treads
(Ticket Slot Assembly Grouted into Position)
3. Glass Cantilevered Canopy
4. Glass Stringer Beams
5. Radiart Panels + Reflector Parts + LEDs
6. Load Bearing Glass Walls
7. Glass Sidewalls
8. Prefabricated Fiberglass Booth
9. Skid Mounted Mechanical Equipment
10. TKTS Counters
11. Raised Form Assembly
12. Geothermal Wells

1. 玻璃围栏
2. 夹层玻璃踏板(自动售票机以石灰浆固定)
3. 玻璃悬臂式屋顶
4. 玻璃纵梁
5. 拉低尔特板+反射部分+LED灯
6. 玻璃侧墙
7. 承重玻璃墙
8. 预制纤维玻璃售票厅
9. 滑轨式安装机械设备
10. TKTS柜台
11. 凸起模块组合
12. 地热井

AXON

Urban Outfitters Corporate Campus
城客服饰公司园区

Jury Comments:
This is a great example of a reuse project. The industrial buildings were beautifully renovated, resulting in an open, collaborative environment that reflects the image of the company that they portray in their clothing stores.

评委评语：
这是一个再利用项目的典范。工业建筑得到了完美的修复，形成了开放的合作型环境，反映了公司的服装店的形象。

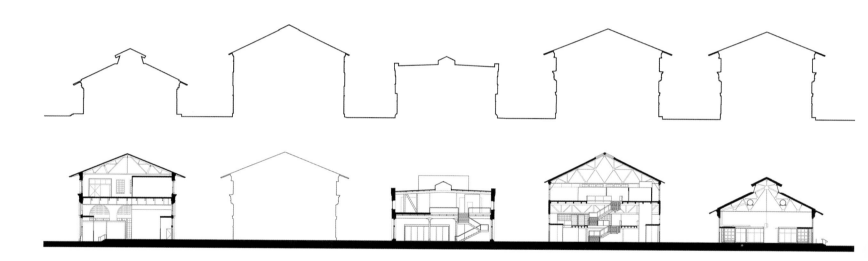

Notes of Interest

The Urban Outfitters Corporate Campus, housed in five rehabilitated buildings in the historic Philadelphia Navy Yard, provides design studio and office space for the company's distinctive retail brands while celebrating the idiosyncratic remnants of more than 125 years of ship-building.

When Urban Outfitters first considered the site, the existing structures were dilapidated. Despite the decay, the soul of the Yard spoke to the company's founder. Each building now houses a different division of the company. Design, documentation, and renovation were completed within 23 months.

The design centers on utilizing the factory characteristics of the buildings – industrial materiality, open volumes and access to daylight – to repurpose the buildings' major function from production to creativity. The synthesis of four measures – art, culture, economy, and environment – results in the transformation from a public, production-based yard to a private, creativity-based one.

Associate Architect: H2L2 Architects Planners, LLC
Consultant: Advanced GeoServices, Jim Larson
Engineer: Paul H. Yeomans, Inc., Meyer, Borgman, and Johnson, Inc.
General Contractor: Blue Rock Construction
Landscape Architect: D.I.R.T. Studio
Owner: Urban Outfitters, Inc.

合作建筑师：H2L2 建筑规划公司
顾问：高级吉奥服务公司、吉姆·拉尔森
工程师：保罗·H·尤门公司、MBJ 公司
总承包商：蓝岩建筑工程公司
景观建筑师：D.I.R.T. 工作室
所有人：城客服饰公司

项目特色

城客服饰公司园区坐落在原费城海军造船厂的五座大楼里，为该公司的特色零售品牌提供了设计工作室和办公空间，同时也对拥有125年历史的造船厂显示了应有的敬意。

城客服饰公司首先考虑了场地，原有的结构已经被荒废。尽管已经破败不堪，造船厂的灵魂仍然打动了公司的创始人。每座大楼内部都设有公司不同的部门。设计、存档和修复工作在23个月的时间内完成。

设计以利用建筑的工厂特色为中心——工业材料、开放空间、阳光充足，将建筑的功能从生产转为创造。四种措施——艺术、文化、经济、环境的综合——实现了公共生产工厂到私密创意工作室的转变。

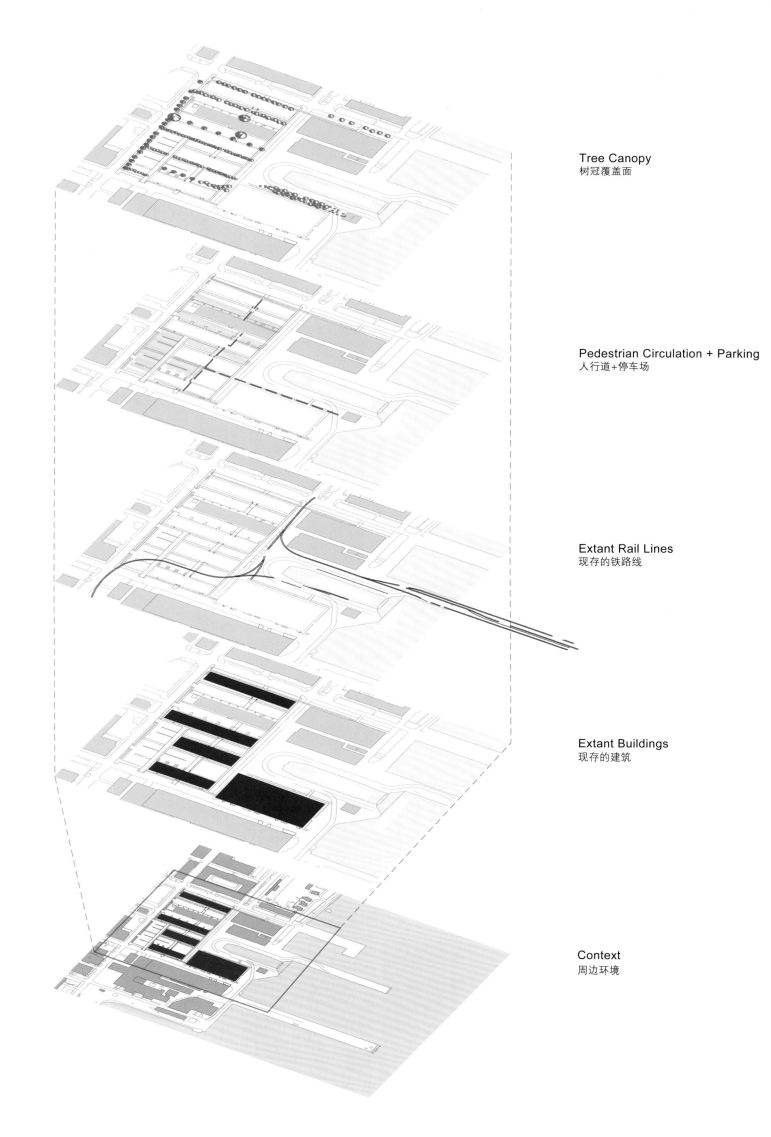

081

Yale University Art Gallery, Kahn Building Renovation

耶鲁大学美术馆——卡恩楼翻新

Jury Comments:
The architectural clarity of the building had been greatly compromised over the years... The original intent is now, once again, perfectly clear... A masterful response to the work of a master.

评委评语：
建筑的通透感在成年累月中被大大地消减。翻修让原有的设计意图重获新生。设计师对大师作品的改造纯熟而巧妙。

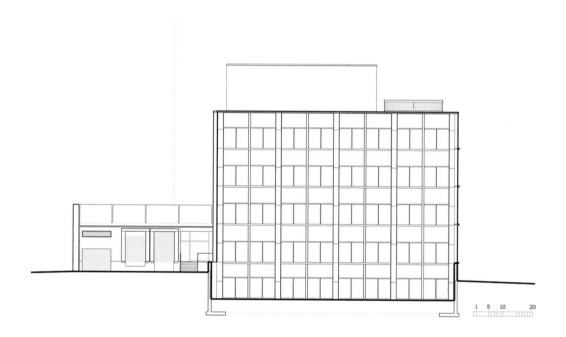

Notes of Interest

Completed in 1953, the Kahn Building is widely considered to be the visionary American architect's first masterpiece and a significant turning point in the history of American museum architecture. Constructed of masonry, glass, and steel, the building has been acclaimed for the bold geometry of its design, its daring use of space and light, and its technical innovations.

The renovation of the landmark building reestablishes its original purity and integrity, restoring many original design features that had become altered or obscured over the years. Roofed over in the 1970's to create additional gallery space, an exterior courtyard has been restored as an open exterior sculpture space. On the second and third floors, extraneous partitions have been removed, and the individual galleries are now revealed in spacious, unobstructed vistas according to Kahn's original vision.

Consultant: Wolf & Company, Robert Schwartz & Associates, Shen Milsom & Wilke, Hughes Associates
Engineer: Robert Silman Associates PC, Altieri Sebor Wieber Consulting Engineers, David DeLong
General Contractor: Barr & Barr Builders
Landscape Architect: Towers | Golde
Owner: Yale University

项目特色

卡恩楼建于1953年，被广泛认为是视觉系美国建筑师的首个杰作，也是美国博物馆建筑历史上的转折点。建筑由石材、玻璃和钢铁构造而成，以其大胆的几何图形设计、空间和光线运用以及技术创新而闻名。

这座地标式建筑的翻新工程重塑了其原有的纯粹和完整感，恢复了许多原有的设计特色。露天庭院经过修复形成了一个雕塑花园。二楼和三楼的附加隔断都拆除，令独立画廊呈现出卡恩楼最初设计的开放感和无障碍视野感。

顾问： 沃夫公司、罗伯特·施瓦兹事务所、沈·米尔索姆&威尔克、休斯事务所
工程师： 罗伯特·希尔曼工程事务所、阿尔蒂里·瑟博尔·韦伯工程咨询公司、大卫·德隆
总承包商： 巴尔&巴尔建筑公司
景观建筑师： 陶尔斯|戈尔德
所有人： 耶鲁大学

Architect / 建筑师
Polshek Partnership Architects
波尔舍克建筑事务所

Location / 项目地点
New Haven
纽黑文

Photo Credit / 图片版权
© Elizabeth Felicella
伊利莎白·菲利斯拉

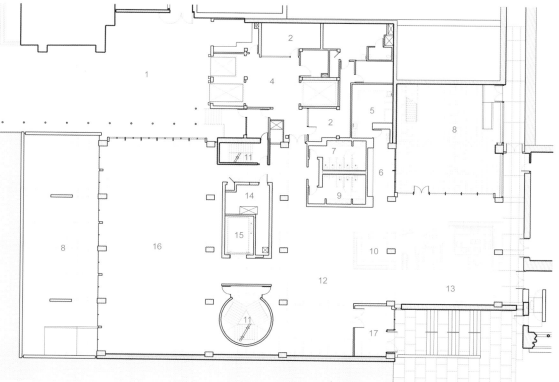

For Reference Only
Site & Ground Floor Plan:
1. Driveway
2. Admin.
3. York st.
4. Loading dock
5. Catering
6. Lockers
7. Women
8. Court
9. Men
10. Reception
11. Stair
12. Lobby
13. Media lounge
14. Mech.
15. Elev.
16. Temporary exhibition gallery
17. Vest.
18. Chapel st.

仅供参考
场地和地面层平面图：
1. 车道
2. 行政办公室
3. 约克街
4. 卸货码头
5. 餐厅
6. 更衣室
7. 女洗手间
8. 庭院
9. 男洗手间
10. 前台
11. 楼梯
12. 大厅
13. 媒体休息室
14. 机械室
15. 电梯
16. 临时展览厅
17. 门廊
18. 教堂街

085

CHANEL Robertson Boulevard
香奈儿罗伯森大道店

Jury Comments:
What great restraint-elegantly understated; the content is beautifully executed.

评委评语：
完美的约束——优雅而低调；整体环境华丽美观。

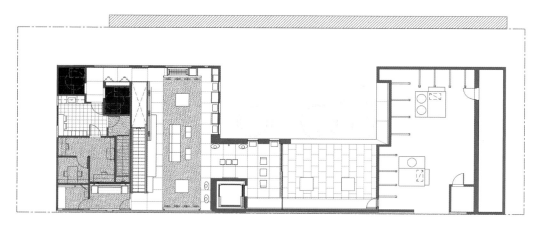

Notes of Interest

The monolithic white plaster façade above the entry references Chelsea galleries as do the clean, spare finishes within. Zoning restrictions dictated preservation of an existing building, which was stripped down to the bare wood frame.

Through the open street façade a gently rising promenade passes through three distinct "zones" to a semi obscured stair hinting at continued exploration above. The entrance, with its high glossy black and white barrisol ceilings, corresponding black and white polished stone floors and stage lighting, is exuberantly theatrical. Next, a series of lighted coves running up one wall, across the ceiling, and down the opposite wall orients the customer to the courtyard and the Southern California light.

The U-shaped first floor is organized around an exterior courtyard with a plaster façade punctured by 17 uniformly sized openings. The unifying courtyard is present in each "room", but always freshly orientated in both plan and section.

Engineer: Murphy Burr Curry, Rosini Engineering
General Contractor: Dickinson Cameron Construction
Owner: CHANEL
顾问：墨菲·布尔·克里、罗西尼工程公司
总承包商：迪金森·卡梅隆建筑公司
所有人：香奈儿集团

项目特色

入口上方纯粹的白色石膏外立面参考了切尔西美术馆的风格，简洁干净，将所有装饰留在了室内。区域限制令要求设计保留仅剩裸露木结构的原有建筑。

从街面进入，一条缓缓上升的走道穿过三个独立区域直达半层楼梯，引导顾客继续探索。入口的高光黑白巴力天花板与黑白色抛光石地面相互呼应，极富戏剧效果。一系列凹灯在紧邻的墙面上呈上升趋势排列，然后又在对墙下降，引导着顾客走入庭院之中。

U形一层结构围绕着露天庭院展开，石膏外立面上共有17个大小相同的窗口。每个"房间"都能看到庭院的景象，但是视野角度又不尽相同。

Architect /建筑师	Location /项目地点	Photo Credit /图片版权
Peter Marino Architect	Los Angeles, California	© Paul Warchol Photography
彼得·马里诺建筑事务所	加利福尼亚州,洛杉矶	保罗·瓦克尔摄影

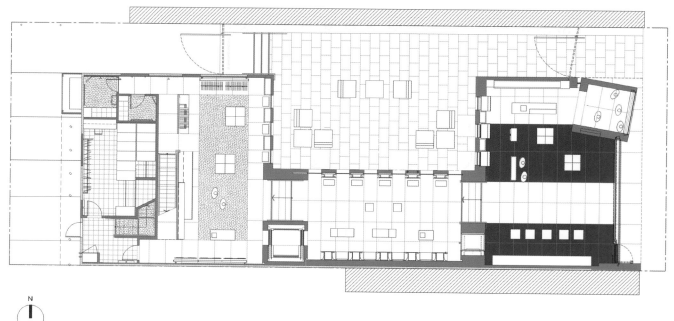

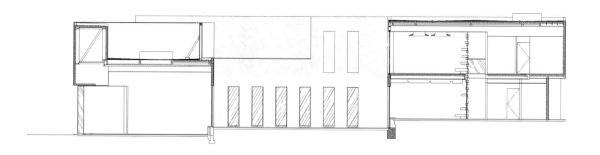

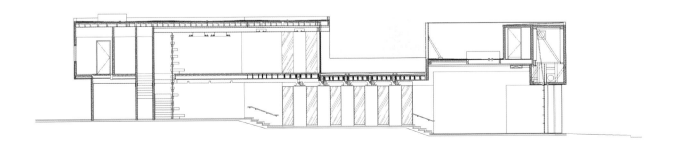

Craftsteak

工艺牛排店

Jury Comments:
It very successfully takes a large volume of space and makes it comfortable. It is lovingly crafted with attention to detail which brings the scale down by humans for humans.

评委评语:
项目成功地将一个大型空间改造得舒适宜人,十分注重细部工艺,坚持以人为本。

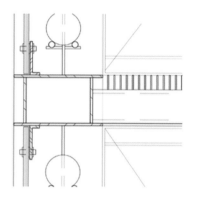 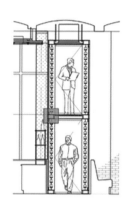 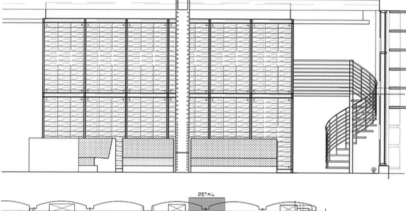

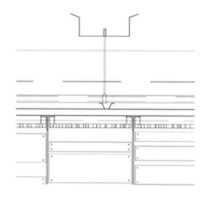 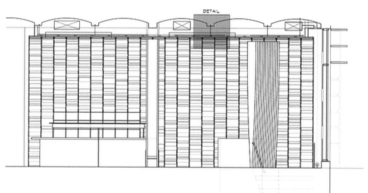

Notes of Interest

The architect's ultimate goal for the restaurant project was to shape, within the one hundred year old shell of this previous National Biscuit Company bakery building, a simple yet texturally and spatially rich interior that integrates the context with the food service both functionally and metaphorically.

Inspired by chef and owner Tom Colicchio's culinary approach of uncomplicated respect for the ingredient, all furnishings and fittings, such as the walnut and steel dining tables, were designed to celebrate their materials and the simple craftsmanship used to assemble them. In addition to this self-assigned goal, the architect also accommodated Colicchio's desire for 225 total seats, 2,000-bottle wine storage, and 3,000-square-foot kitchen, contained within a 3,500-square-foot first floor and a 4,500-square-foot cellar.

Consultant: Alliance Food Equipment Corp., Archetype Consultants Inc.
Engineer: Koutsoubis, Alonso Associates, P.E., P.C., AMA Consulting Engineers, P.C.
General Contractor: MG & Company
Owner: FoodCraft LLC

顾问: 联盟食品设备公司、原型咨询公司
工程师: 库特苏比斯·阿隆索事务所、AMA工程咨询公司
总承包商: MG公司
所有人: 食品工艺公司

Architect / 建筑师	Location / 项目地点	Photo Credit / 图片版权
Bentel & Bentel Architects 本特尔&本特尔建筑事务所	New York City, New York 纽约州，纽约	© Eduard Hueber 爱德华·修伯尔

项目特色

建筑师的终极目标是在拥有百年历史的前国家饼干公司面包坊内打造简单而质地丰富的室内设计,让环境与食品服务在功能上和象征意义上完美结合。

设计受到了主厨和店长汤姆·科利基奥的简单食材烹饪手法的启发,所有装饰和配件(例如胡桃木和钢铁制成的餐桌)都以材料为基础,采用了最简单的工艺进行组合。

除了这个自有目标之外,建筑师还根据科利基奥的需求,打造了225个坐席、能够收藏2,000瓶红酒的藏酒架、一个约280平方米的大厨房。所有设施都设在325平方米的一楼空间和418平方米的地下室中。

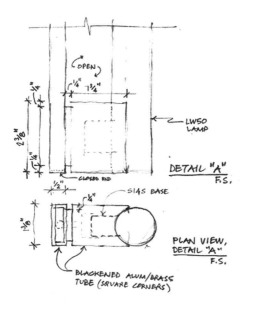
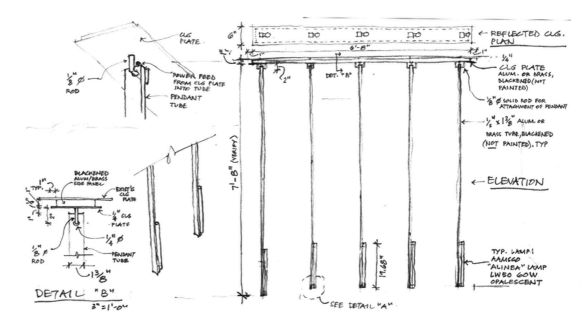

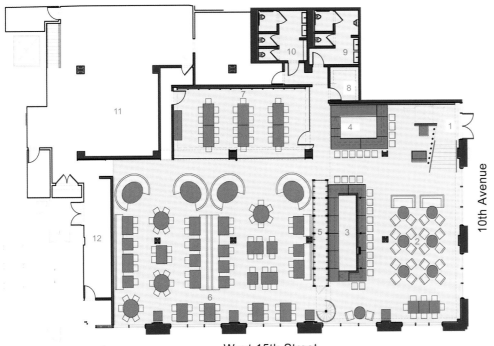

Plan:
1. Entry
2. Lounge
3. Bar
4. Raw bar
5. Wine vault
6. Main dining area
7. Private dining room
8. Coats
9. Men
10. Women
11. Kitchen
12. Service station

平面图：
1. 入口
2. 休息室
3. 吧台
4. 生鲜吧
5. 酒架
6. 主就餐区
7. 私人包间
8. 衣帽间
9. 男洗手间
10. 女洗手间
11. 厨房
12. 服务台

Data
达塔公司

Jury Comments:
The project is well executed with a certain level of craft; the details are nice, clean, and edgy, yet a fun work environment.

评委评语：
项目采用了精湛的工艺，细部精美、简洁而前卫，同时又是一个充满趣味的工作环境。

Notes of Interest

The client is one of America's leading providers of mailing lists, marketing data, sales leads, and research data. The client's challenge to the architect was to create a fresh new design for their office that expresses who they are.

The organizational strategy was to create a centrally located circulation core that sets the mood for the office and connects all of the spaces together. The design energy and most of the $28 per square foot budget was focused on three elements: an etched glass conference room wall expressing the company's data, a cut and bent wall/ceiling form which connects the office together, and galvanized metal shed wall panels to express both ideas of technology and the Midwest rural vernacular of the company's founding location.

Engineer: Alvine and Associates, Canelli Engineering, Inc.
General Contractor: KSI Construction
Owner: US Data

工程师： 阿尔维恩事务所、卡内利工程公司
总承包商： KSI建筑公司
所有人： 美国达塔公司

项目特色

项目的委托人是美国顶尖数据供应商，专门提供邮件列表、市场数据、销售渠道和研究数据。委托人要求建筑师为公司打造一个全新的设计办公区，设计要体现公司的特色。

项目的组织结构策略是围绕着一个中心交通核奠定工作氛围，同时也将各个空间连接在一起。设计资源和预算主要聚焦于三个方面：展示公司数据的磨砂玻璃会议室墙面、连接办公空间的切割弧面墙面/天花板造型以及展示技术理念和公司发源地（美国中西部）风情的镀锌金属墙板。

Architect / 建筑师	Location / 项目地点	Photo Credit / 图片版权
Randy Brown Architects 兰迪·布朗建筑事务所	Omaha 奥马哈	© Assassi 2009 2009事务所

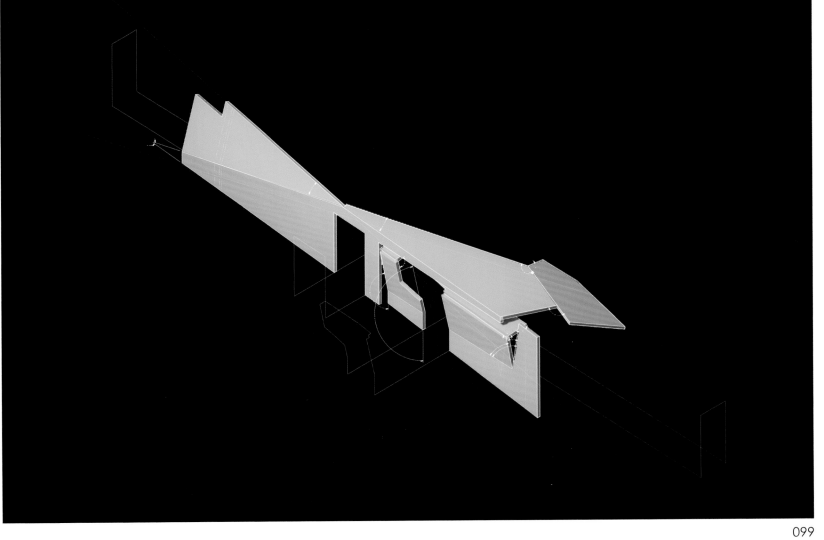

Exeter Schools Multipurpose Space
埃克塞特学校多功能空间

Jury Comments:
Such a joyful, bountiful amount of architecture, achieved under constraints of site and budget... turning a small budget into a functional and beautiful place is outstanding to see.

评委评语:
项目在有限的场地和预算内打造了一座令人愉悦的宽敞空间，将低预算成功地转变为兼具功能性和美感的优秀建筑。

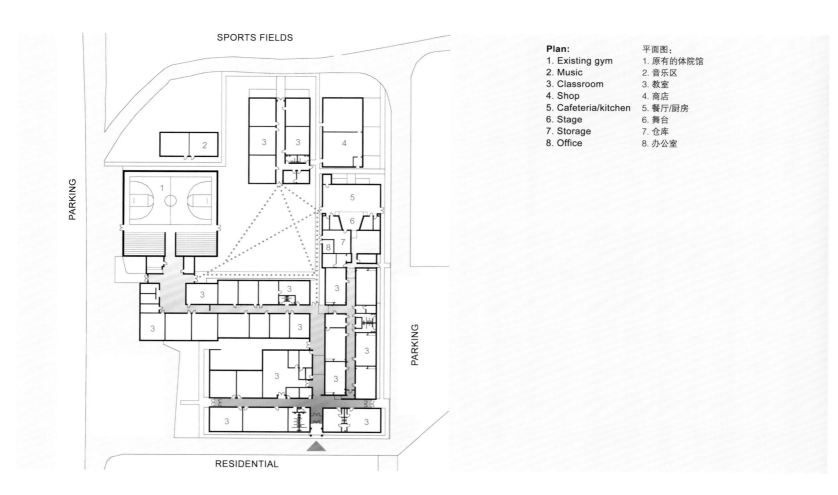

Plan:
1. Existing gym
2. Music
3. Classroom
4. Shop
5. Cafeteria/kitchen
6. Stage
7. Storage
8. Office

平面图:
1. 原有的体院馆
2. 音乐区
3. 教室
4. 商店
5. 餐厅/厨房
6. 舞台
7. 仓库
8. 办公室

Notes of Interest
The challenge was to design a single space that functions well as a cafeteria, practice gym and performance hall for a rural K-12 school district on a limited budget.
Solving the acoustic challenges of these varied uses led to a solution derived from a sushi roll – absorptive on its outermost layer with a thin, reflective inner layer. The solution is executed as a simple steel frame structure inserted into an existing courtyard with an outer layer of Tectum and an inner layer of perforated wood panels. Attention is paid to detailing the wood panels to distribute sound appropriately for performances while protecting light fixtures and mechanical systems for use as a gymnasium.

Consultant: Environmental Market Solutions, Inc., Bruce Moore, AIA
Engineer: Jones & Associates Structural Engineers, Genesis Mechanical
General Contractor: Springfield Builders, Inc.
Owner: Exeter R-VI School District

顾问: 环境市场方案公司、布鲁斯·摩尔（美国建筑师协会）
工程师: 琼斯结构工程事务所、詹尼西斯机械
总承包商: 斯普林菲尔德建筑公司
所有人: 埃克塞特R-VI学区

Architect / 建筑师	Location / 项目地点	Photo Credit / 图片版权
Dake Wells Architecture 德克·威尔斯建筑事务所	Exeter, Missouri 密苏里，埃克塞特	© Gayle Babcock; Architectural Imageworks, LLC 盖尔·巴布科克、建筑图像公司

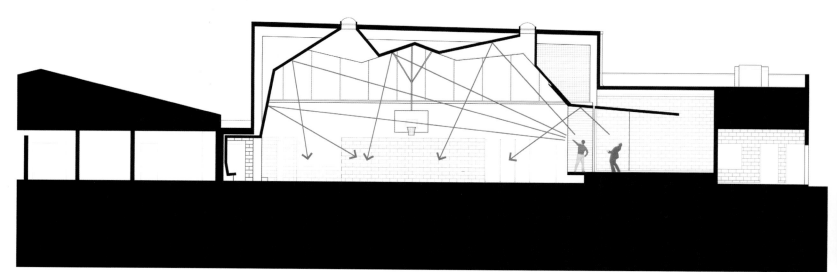

项目特色

设计要求在有限的预算内为一个乡村K-12（从幼儿园到十二年级）学区打造兼具食堂、体育馆和表演厅的单一空间。

多功能空间的音响效果问题被一个类似寿司卷的方案所解决——墙面的最外层可以吸音，内层轻薄而具有反射性。一个简单的钢框架结构嵌入了原有的庭院，外层结构为致密层，内层则是空心板。木板的细部设计既有利于演出的音效扩散，又能在运动时保护照明设施和机械系统。

Historic Central Park West Residence
中央公园西部住宅

Jury Comments:
Historic elements are re-detailed with the white panels, allowing them to simultaneously exist and recede, and the dialogue with the modern elements of dark wood reinvigorates the architecture.

评委评语:
历史元素通过白色的面板得到了重新体现,使存在感和弱化感并存,并且与振兴建筑的黑木现代元素形成了对话。

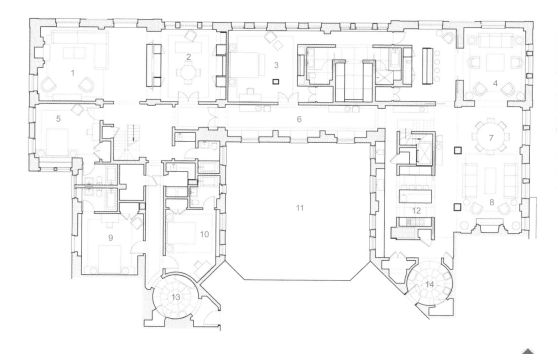

12th Floor Plan:
1. Guest living room
2. Study
3. Master bedroom
4. Sitting room
5. Guest bedroom 1
6. Gallery
7. Dinning room
8. Living room 1
9. Guest bedroom 2
10. Guest bedroom 3
11. Courtyard
12. Kitchen
13. Entry 2
14. Entry 1

12层平面图:
1. 客用起居室
2. 书房
3. 主卧室
4. 客厅
5. 客卧1
6. 走廊
7. 餐厅
8. 起居室1
9. 客卧2
10. 客卧3
11. 庭院
12. 厨房
13. 入口1
14. 入口2

Notes of Interest

This project called for combining two untouched, disparate penthouses (circa 1920) in one of Manhattan's noted landmark beaux-arts revival buildings to create one cohesive, seamless residence. It had to retain the best of the historic past, while still being appropriate to our time.

Additional goals involved taking full advantage of the four exposures of light, mezzanine, conservatory, rooftop access and views of Manhattan's Central Park. In addition, the architect provided the philanthropist owner with a residence easily maneuvered and divided into "public" and "private" spaces for work and family.

Consultant: Schwinghammer Lighting
Engineer: I.P. Group, Ross Dalland, P.E.
General Contractor: 3-D Laboratory, Inc
Landscape Architect: R/F Landscape Architecture P.C.
Owner: Jon Stryker

工程师:I.P.集团、罗斯·达兰德
总承包商:3-D实验室
景观建筑师:R/F景观建筑事务所
所有人:乔恩·斯特瑞克

项目特色

项目要求将一座曼哈顿知名艺术复兴建筑中的两所截然不同的顶层公寓(约建于1920年)合并为一个连续的住宅空间。在保持公寓的历史价值的同时,项目还必须适应现代生活需求。

附加目标包括充分利用四面透光性、夹层楼板、温室、屋顶通道和曼哈顿中央公园的景色等优势。此外,建筑师还给身为慈善家的所有人提供了一个便于简单改造的空间。住宅内"公""私"空间的划分让工作和生活有机的结合在一起。

Architect /建筑师
Shelton, Mindel & Associates
谢尔顿和民德尔事务所

Location /项目地点
New York City, New York
纽约州，纽约

Photo Credit /图片版权
© Michael Moran Photography
迈克尔·莫兰摄影

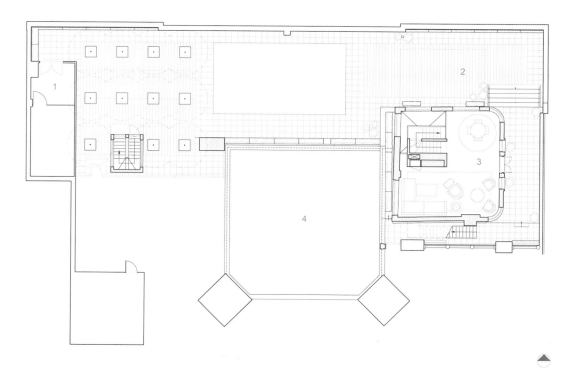

Penthouse Floor Plan:
1. Utlts Utilities
2. Terrace
3. Media Room
4. Courtyard

顶层平面图:
1. 公共房
2. 平台
3. 媒体室
4. 庭院

The Cathedral of Christ the Light

耶稣光明大教堂

Jury Comments:
The project exhibits some of the most original rethinking of architectural enclosure and form since the Gothic period.

评委评语：
该项目展示了自哥特时期以来对建筑结构和造型的最原始的反思。

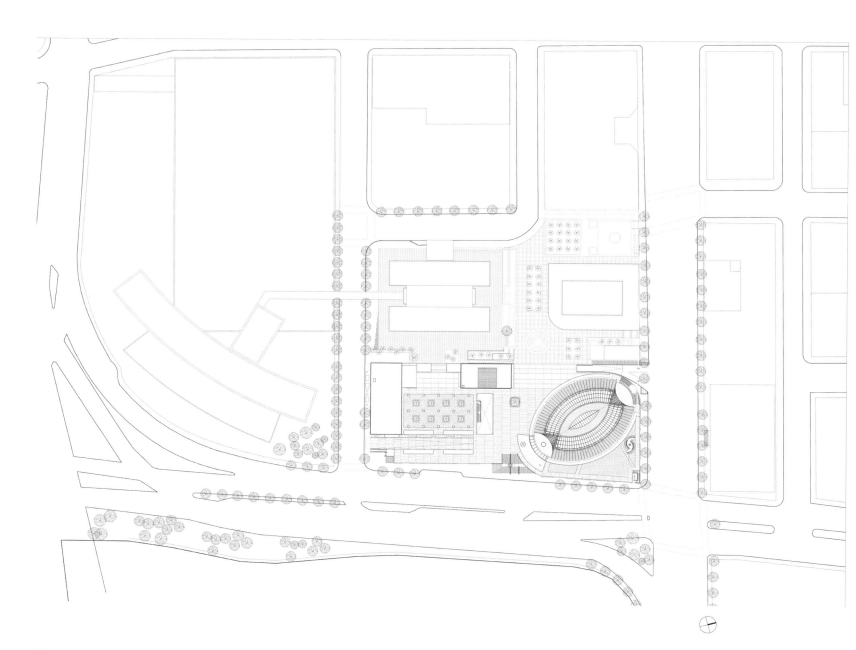

Notes of Interest

The Diocese challenged the design team to create a building for the ages. As a result, the 1,350-seat sanctuary, with its side chapels, baptistery, health and legal clinics and dependencies, will honor its religious and civic obligations to the Catholic Diocese and the city for centuries to come.

Through its poetic introduction, indirect daylight ennobles modest materials-primarily wood, glass and concrete. Triangular aluminum panels form the petal-shaped Alpha Window, which diffuses light 100 feet above the Cathedral's entrance. The Omega Window resonates with the surrounding structure metaphorically and physically through its experimental use of light, re-imagining a 12th-century depiction of Christ from the façade of Chartres Cathedral in France through over 94,000 pixels cut into the Window's triangular aluminum panels.

Associate Architect: Kendall/Heaton Associates, Inc.
Consultant: Conversion Management Associates, Inc., The Engineering Enterprise, Claude R. Engle Lighting Consultants, Shen Milsom & Wilke, Inc.
Engineer: Skidmore, Owings & Merrill LLP, Taylor Engineering
General Contractor: Webcor Builders, Oliver + Co.
Landscape Architect: Peter Walker and Partners
Owner: Roman Catholic Diocese of Oakland

顾问： 变换管理公司、工程企业、克劳德·R·恩格尔照明顾问公司、沈·米尔萨姆＆威尔克公司
工程师： SOM公司、泰勒工程公司
总承包商： 韦伯柯尔建筑公司、奥利弗公司
景观建筑师： 彼得·怀特事务所
所有人： 罗马天主教奥克兰教区

Architect
Skidmore, Owings & Merrill LLP
SOM 公司

Location
Oakland, California
加利福尼亚州，奥克兰

Photo Credit
© Cesar Rubio, © Timothy Hursley
All drawing are © SOM
凯撒·卢比奥、狄默思·赫斯利、SOM 公司（所有技术图）

© Cesar Rubio

项目特色

教会要求设计团队打造一座经得起时间考验的建筑。因此,他们打造了拥有 1,350 个坐席的圣殿、附属小礼拜堂、洗礼堂、保健和法律诊所和相关配套设施,充分体现了天主教教区和城市的宗教和社会职责。

在充满诗意的入口,间接的阳光让低调的材料(主要是木材、玻璃和混凝土)显得高贵。三角形铝板组成了花瓣造型的阿尔法窗(前窗),阳光透过窗口漫射到教堂的入口。欧米茄窗(后窗)与周边的结构通过光线的运用相互辉映。窗口三角形铝板上的 94,000 个像素点重现了 12 世纪法国沙特尔大教堂外立面上的耶稣像。

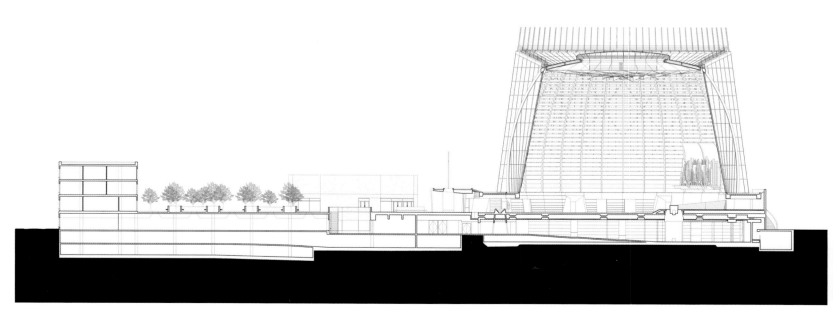

© Timothy Hursley

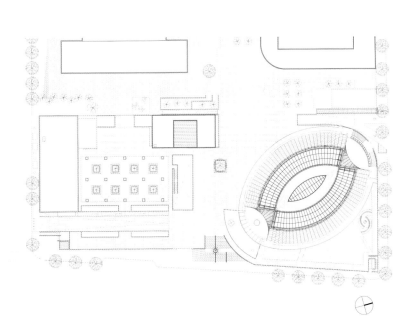
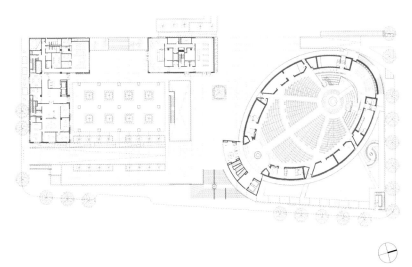

111

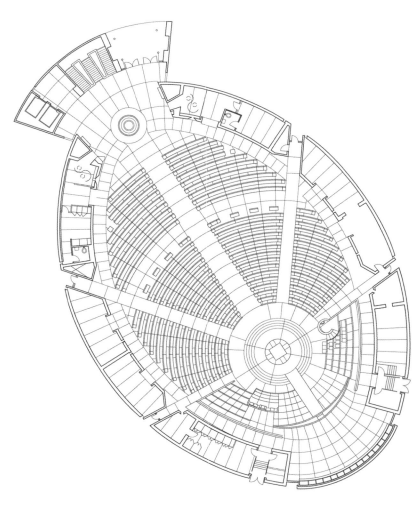

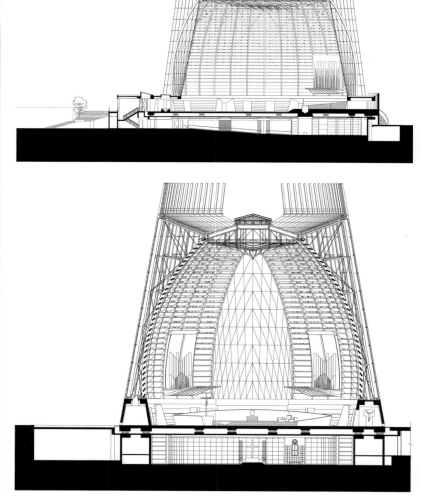

Vera Wang Boutique
王薇薇精品店

Jury Comments:
It is all about the subtlety – of the details that are very, very refined. You don't see them at first and they reveal themselves as you move within the space.

评委评语：
项目极尽精致，每个细节都无比精美。或许你一眼看不出来，但随着你在空间内的走动，它们将会逐一浮现。

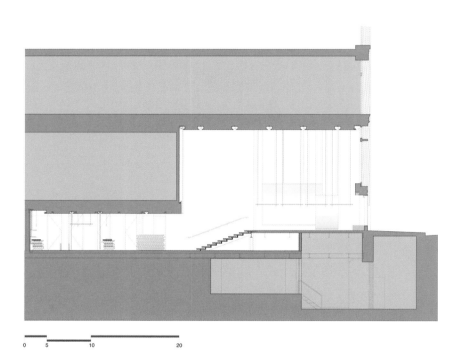
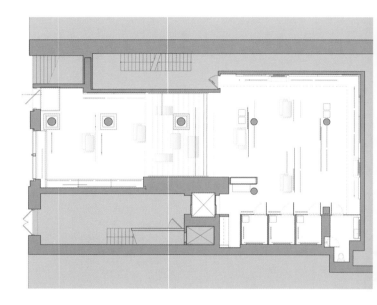

Notes of Interest
Customers enter as performers on a stage, stepping into the elevated, double-height proscenium at the front of the store. The spatial sequence unfolds down a full-width, white Corian grand stair, which transitions into the more intimate display and changing area at the rear of the space.
With LED backlighting, the steps appear to float; they double as seating for special events or a display riser with translucent acrylic platforms.
Reflecting the juxtapositions that characterize Vera Wang's fashion design, the material palette is based on a series of complementary contrasts. High-gloss, white epoxy flooring contrasts with diaphanous scrims; synthetic acrylic partitions feature hand-sculpted, curved edges, while natural architectural plaster is polished to the point of abstraction. Lighting is as much a physical material as the plaster and steel, and is used to simulate a range of atmospheric states.

Consultant: Tillotson Design Associates
Engineer: Robert Silman Associates, Edwards & Zuck Consulting Engineers
General Contractor: Michilli, Inc.
Landscape Architect: R/F Landscape Architecture P.C.
Owner: Vera Wang Group
顾问：蒂尔洛森设计事务所
工程师：罗伯特·希尔曼事务所、爱德华兹 & 扎克工程咨询公司
总承包商：米奇里公司
景观建筑师：R/F 景观建筑事务所
所有人：王薇薇集团

Architect
Gabellini Sheppard Associates
加布里尼·史柏德建筑事务所

Location
New York City, New York
纽约州，纽约

Photo Credit
© Paul Warchol
保罗·瓦尔孔

项目特色

走进精品店的顾客仿佛是走上舞台的表演者，他们登上了店内前方高高的舞台。空间序列随着宽阔的可丽耐人造大理石楼梯展开，引领人们进入更私密的展览区和店铺后方的更衣室。

LED 背景光让台阶看起来好像悬浮一样；它们在特殊活动时可以提供座位或是展示平台。

材料的搭配以一系列相辅相成的对比为基础，反映了王薇薇时装设计的特色。高光白色树脂地板与透明的纱帘形成了对比；人造亚克力隔断以手工雕刻的边缘为特色；天然建筑石膏则被抛光形成抽象的角度。照明与石膏和钢材一样，都是基础材料，为空间带来了各具特色的不同氛围。

A Civic Vision for the Central Delaware River
德拉瓦河中游市政前景

Jury Comments:
Focusing on seven miles of Philadelphia's former industrial waterfront, this concept for reclaiming severely dilapidated real estate is long overdue. This is a very appropriate set of solutions to a set of longstanding problems and a sustainable approach to the re-invigoration of existing facilities.

评委评语:
项目聚焦于费城7英里长的工业滨水区,恢复了长期以来遭受严重破坏的房地产项目。对于长期存在的问题来说,这是一系列值得欣赏的解决方案;它通过可持续方式复兴了现有设施。

Notes of Interest

As the lead design consultant for this mayoral initiative, the firm created a new vision for seven miles of the Delaware River in Philadelphia. Currently cut off from the city by the intrusion of I-95, this riverfront is comprised of underutilized post-industrial land and big-box development, and is subject to unregulated residential speculation. The plan emphasizes the ecological and economic value of the waterfront and sets forth a framework that the city can follow to generate new, cohesive, and sustainable development. This new growth will be organized around parks and open space, providing access to the river and a new movement system, including the decking-over of I-95 and a grand civic boulevard complete with public transit. For the ability of the plan to accommodate the future needs of the city and its people, this project has received numerous endorsements.

Consultant: Penn Project on Civic Engagement, Philadelphia City Planning Commission
Owner: PennPraxis, University of Pennsylvania
顾问: 宾夕法尼亚大学市政管理项目部、费城城市规划委员会
所有人: 宾夕法尼亚大学

Architect / 建筑师
Wallace Roberts & Todd, LLC
华莱士·罗伯斯&陶德公司

Location / 项目地点
Philadelphia, Pennsylvania
宾夕法尼亚州，费城

Photo Credit / 图片版权
© Wallace Roberts & Todd
华莱士·罗伯斯&陶德公司

项目特色

作为这个市政项目的首席设计顾问,华莱士·罗伯斯&陶德公司为德拉瓦河在费城境内的7英里区域打造了新前景。目前,河畔区域在城内被I-95公路所隔断,由未被充分利用的后工业用地和大方盒式开发项目组成,并且受到了未经调整的住宅投资项目的影响。规划强调了滨水区域的生态和经济价值,提出了一个城市能够依照生成全新的可持续开发的框架。这个新开发将围绕公园和开放空间展开,提供通往河流的入口和全新的移动系统,其中包括在I-95公路上搭建桥面板和一条宏伟的城市大道。规划在满足未来城市和民众需求方面获得了众多赞赏。

PARKS AND OPEN SPACE
公园和开放空间

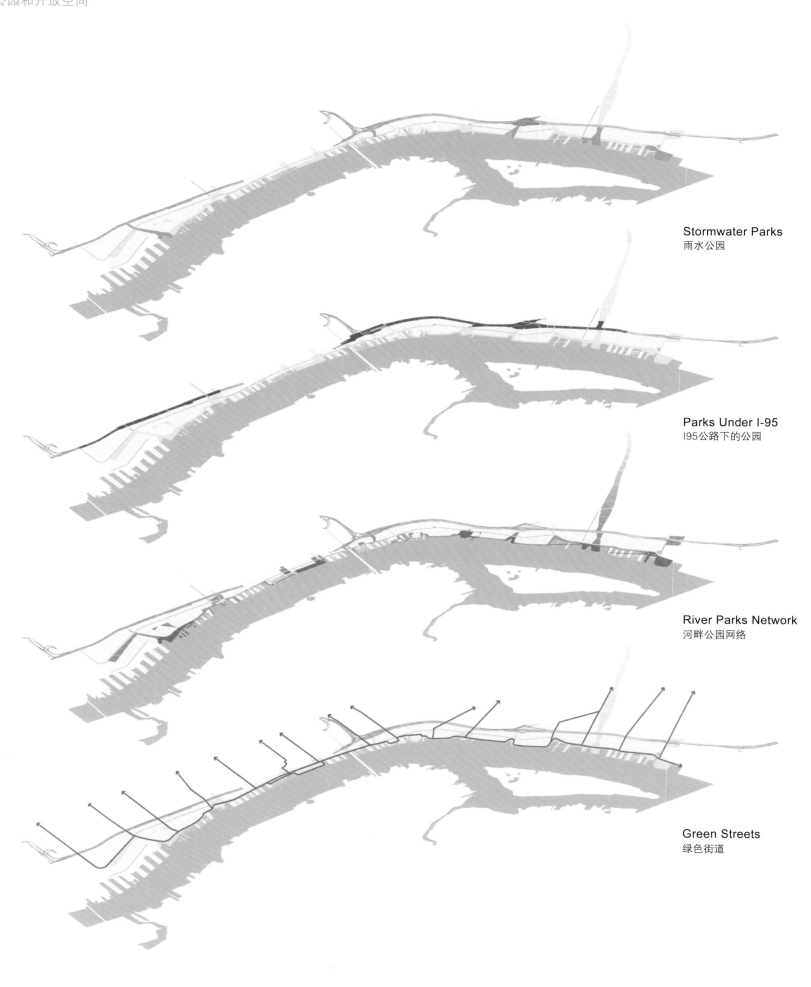

Stormwater Parks
雨水公园

Parks Under I-95
I95公路下的公园

River Parks Network
河畔公园网络

Green Streets
绿色街道

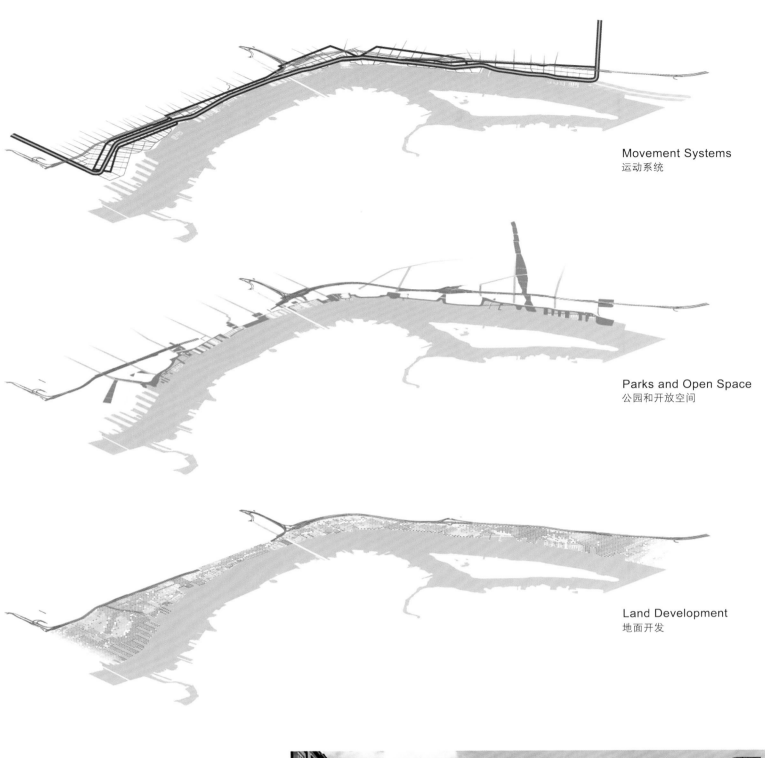

Movement Systems
运动系统

Parks and Open Space
公园和开放空间

Land Development
地面开发

Connections:
MacArthur Park District Master Plan
连接：麦克阿瑟公园区整体规划

Jury Comments:
Connecting a historically residential area of Little Rock in a way that also extends the park beyond its highway-induced isolation, this is an excellent endeavor to mitigate the effects of other-scaled urban infrastructures and use connective landscape amenities to enhance the quality of the urban experience and "eyes on-the-street" security.

评委评语：
项目连接了小石城的一个历史住宅区，同时也扩建了被高速公路阻断的公园空间，出色地缓和了城市基本结构所造成的影响并利用连续的景观设施提升了城市体验和街面质量。

Notes of Interest
Like waterfronts and transit stops, parks leverage value in urban areas. While much recent attention has been given to the signature mega-park, the value of the small-scale neighborhood park in reinventing the city has been overlooked. Once connecting neighborhoods of differing character, and sponsoring more than 80 residential structures along its edges, the historic MacArthur Park at the edge of downtown Little Rock is radically underutilized as an urban neighborhood asset. Severed from its neighborhoods along two edges by interstate construction in the 1960s, this moribund 40-acre municipal park is left with only 16 residential structures along its frontage. The planning concept optimizes the park's latent economic, environmental, and social potential through improvements to the district's neighborhood infrastructure, enhancing the delivery of ecological and urban services. This counters the greatest ongoing threat to MacArthur Park District's irreplaceable legacy – incompatible low-density, suburban-type development that fails to define street edges, and is inherently cynical of the city. The planning goal is to align the park's capacity to sponsor denser and higher quality mixed-use housing fabric throughout the district with improvements to the park grounds. Rather than treat MacArthur Park as a discrete project, planning for the district's four neighborhoods extends the park's landscape into a larger urban landscape network with MacArthur Park as the anchor.

Associate Architect: University of Arkansas Community Design Center
Consultant: University of Arkansas Little Rock Urban Studies, Donjek Public Finance
Engineer: McClelland Consulting Engineers
Landscape Architect: Oslund and Associates
Owner: City of Little Rock, Parks and Recreation
助理建筑师： 阿肯色大学社区设计中心
顾问： 阿肯色大学小石城城市研究部、唐约克公共财政公司
工程师： 麦克里兰工程咨询公司
景观建筑师： 欧斯兰德事务所
所有人： 小石城公园及文娱管理部

Architect / 建筑师
Conway+Schulte Architects, PA
康韦+舒尔特建筑事务所

Location / 项目地点
Little Rock
小石城

Photo Credit / 图片版权
© Conway+Schulte Architects
康韦+舒尔特建筑事务所

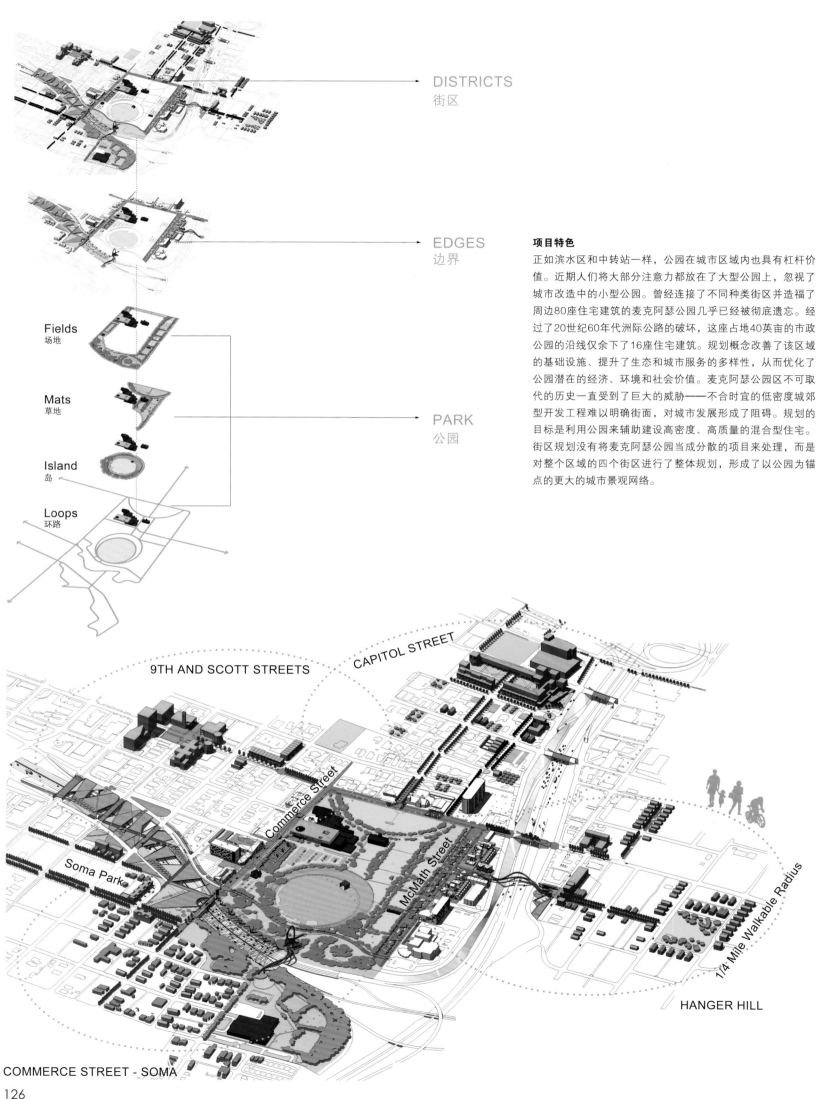

DISTRICTS
街区

EDGES
边界

Fields
场地

Mats
草地

Island
岛

Loops
环路

PARK
公园

项目特色

正如滨水区和中转站一样，公园在城市区域内也具有杠杆价值。近期人们将大部分注意力都放在了大型公园上，忽视了城市改造中的小型公园。曾经连接了不同种类街区并造福了周边80座住宅建筑的麦克阿瑟公园几乎已经被彻底遗忘。经过了20世纪60年代洲际公路的破坏，这座占地40英亩的市政公园的沿线仅余下了16座住宅建筑。规划概念改善了该区域的基础设施、提升了生态和城市服务的多样性，从而优化了公园潜在的经济、环境和社会价值。麦克阿瑟公园区不可取代的历史一直受到了巨大的威胁——不合时宜的低密度城郊型开发工程难以明确街面，对城市发展形成了阻碍。规划的目标是利用公园来辅助建设高密度、高质量的混合型住宅。街区规划没有将麦克阿瑟公园当成分散的项目来处理，而是对整个区域的四个街区进行了整体规划，形成了以公园为锚点的更大的城市景观网络。

COMMERCE STREET - SOMA

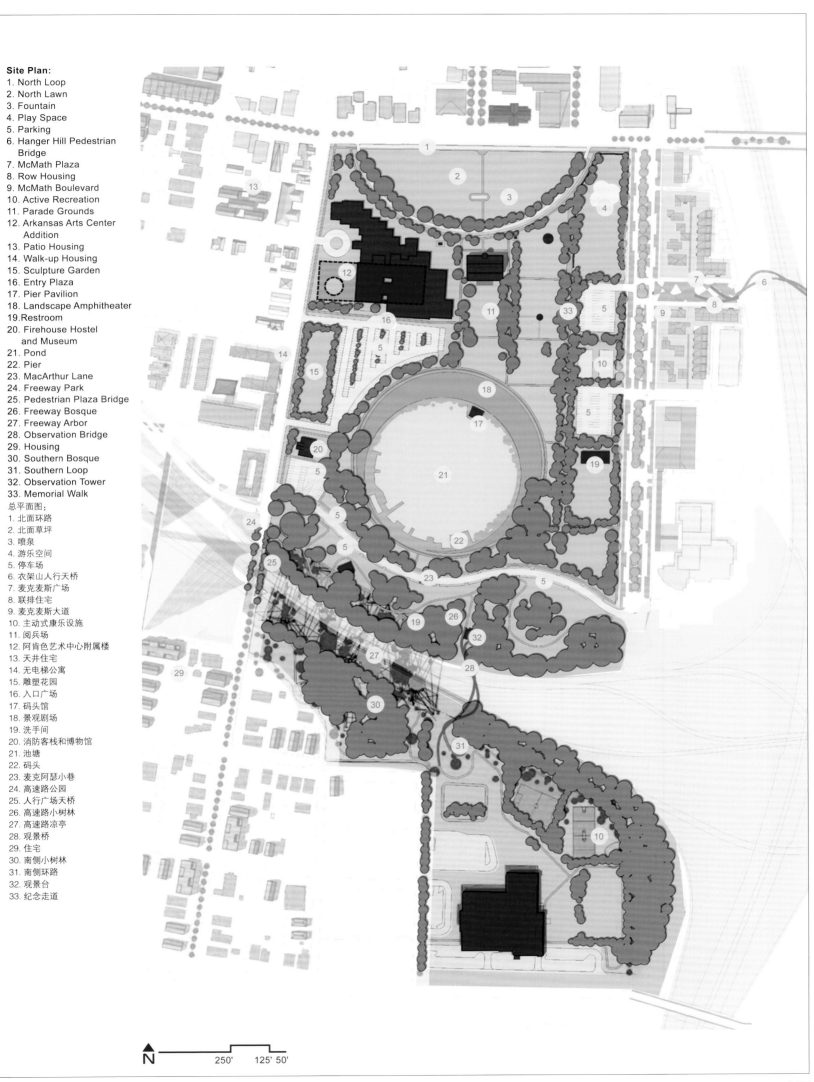

Site Plan:
1. North Loop
2. North Lawn
3. Fountain
4. Play Space
5. Parking
6. Hanger Hill Pedestrian Bridge
7. McMath Plaza
8. Row Housing
9. McMath Boulevard
10. Active Recreation
11. Parade Grounds
12. Arkansas Arts Center Addition
13. Patio Housing
14. Walk-up Housing
15. Sculpture Garden
16. Entry Plaza
17. Pier Pavilion
18. Landscape Amphitheater
19. Restroom
20. Firehouse Hostel and Museum
21. Pond
22. Pier
23. MacArthur Lane
24. Freeway Park
25. Pedestrian Plaza Bridge
26. Freeway Bosque
27. Freeway Arbor
28. Observation Bridge
29. Housing
30. Southern Bosque
31. Southern Loop
32. Observation Tower
33. Memorial Walk

总平面图：
1. 北面环路
2. 北面草坪
3. 喷泉
4. 游乐空间
5. 停车场
6. 衣架山人行天桥
7. 麦克麦斯广场
8. 联排住宅
9. 麦克麦斯大道
10. 主动式康乐设施
11. 阅兵场
12. 阿肯色艺术中心附属楼
13. 天井住宅
14. 无电梯公寓
15. 雕塑花园
16. 入口广场
17. 码头馆
18. 景观剧场
19. 洗手间
20. 消防客栈和博物馆
21. 池塘
22. 码头
23. 麦克阿瑟小巷
24. 高速路公园
25. 人行广场天桥
26. 高速路小树林
27. 高速路凉亭
28. 观景桥
29. 住宅
30. 南侧小树林
31. 南侧环路
32. 观景台
33. 纪念走道

Greenwich South Strategic Framework
格林威治南区战略框架

Jury Comments:
A critical element to the sustainability of the design is that it invites the public to imagine and participate! The potential of this extraordinary large parcel, essentially forgotten for decades, could result in a reconnection that opens up millions of square feet of developable air rights.

评委评语：
设计的可持续特色之一在于它邀请了民众来想象并参与。这一大片区域的潜力在数十年来一直被忽略，终于可以重新进行空间所有权的开发。

Notes of Interest

The primary goal of a study by the Alliance for Downtown New York was to produce a strategic framework for Greenwich South by establishing a set of key principles and objectives to guide both immediate and long-term growth. The architecture firm developed Five Principles to define a vision for the future of Greenwich South as a dense, reconnected, mixed-use neighborhood and lynchpin for Lower Manhattan. Each principle is comprised of a set of clear objectives to be achieved within these goals. In addition to establishing principles and setting goals, the firm also identified a series of clear opportunities for action – from the subtle, genius and immediate to the huge, radical and visionary – to achieve these goals. The project was highly collaborative, employing a Brain Trust as well as a Design Challenge charrette.

Associate Architect: Beyer Blinder Belle
Consultant: OPEN, Marc Kristal
Owner: Alliance for Downtown New York
合作建筑师：拜尔·布林德·贝拉
顾问：OPEN、马克·克里斯特
所有人：纽约下城联盟

Architect / 建筑师
Architecture Research Office
建筑研究办公室

Location / 项目地点
New York City, New York
纽约州，纽约

Photo Credit / 图片版权
© Architecture Research Office
建筑研究办公室

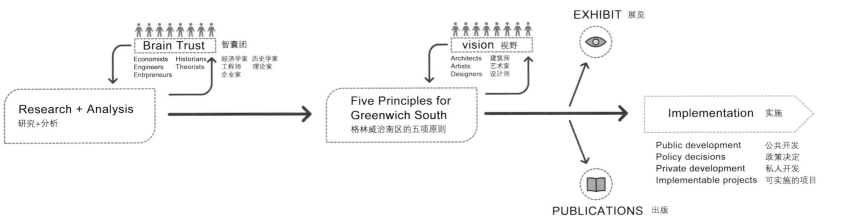

项目特色

纽约下城联盟的主要目标是通过建立一系列指导方针和即时与长期目标而生成格林威治南区的战略框架。建筑公司开发了五项方针将格林威治南区的未来描绘成一个密集的多功能区域，成为下曼哈顿的关键区域。每项方针都有一系列明确的目标要实现。除了建立方针和制定目标，公司还明确了每步行动的——从细微的即时行动到宏观的完整行动——的目标。项目具有高度协作性，采用了智囊团和设计挑战专家研讨会的形式。

Monumental Core Framework Plan
纪念碑核心区框架规划

Jury Comments:
Marking the bicentennial of the original L'Enfant/Ellicott plan and centennial of the McMillan Plan, which created the Mall as we know it today, this phased framework promises to stop degradation of heavily used areas and open less-used venues to greater appreciation and public enjoyment – all within the context of Washington's expanding downtown.

评委评语：
在艾力考特规划二百周年纪念和麦克米伦规划一百周年纪念之际，这个阶段性框架规划承诺在华盛顿不断扩张的市中心内阻止该过度开发区域的退化并为未充分利用区域提供发展机会。

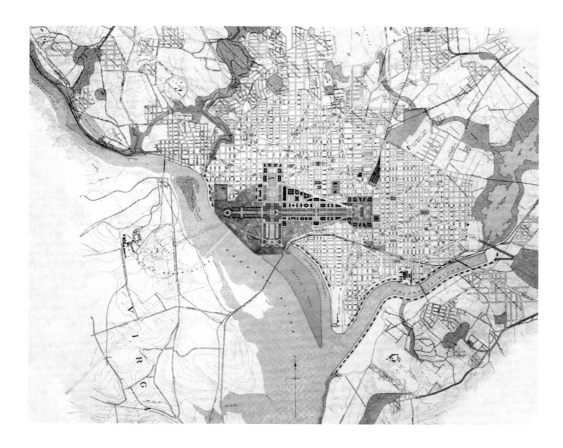

Notes of Interest
The Monumental Core Framework Plan is a proposal sponsored by two federal agencies, the U.S. Commission of Fine Arts and the National Capital Planning Commission, to transform federal precincts surrounding the National Mall into vibrant destinations and to improve connections between the city, the National Mall, and the waterfront. Proceeding from the context of visionary planning for the national capital, the Framework Plan is a practical tool to guide decisions and investment over the next thirty years and complements concurrent planning efforts in Washington.

The Plan proposes a series of sector-by-sector strategies that are designed to protect the National Mall, create distinctive settings for cultural facilities and commemorative works, overcome barriers between the National Mall and the surrounding city, and enhance the monumental core of Washington as a symbolic and sustainable place to work, visit, and live. Initiated to address immediate needs of accommodating national commemorative sites in the monumental core, the Framework Plan also maintains the federal workplace in the central city while providing opportunities for new parks, infrastructure and transportation improvements, and mixed-use public and private development. The Plan is guided by a foundation of best practices to achieve a high quality of urban design, smart growth, and sustainability in strategies for buildings and infrastructure as well as the urban ecological environment.

Associate Architect: EDAW-AECOM
Owner: U.S. Commission of Fine Arts, National Capital Planning Commission
合作建筑师：EDAW–AECOM
所有人：美国国家美术委员会、美国首都规划委员会

Architect / 建筑师	**Location** / 项目地点	**Photo Credit** / 图片版权
U. S. Government 美国政府	Washington, D.C. 华盛顿	©National Capital Planning Commission 美国首都规划委员会

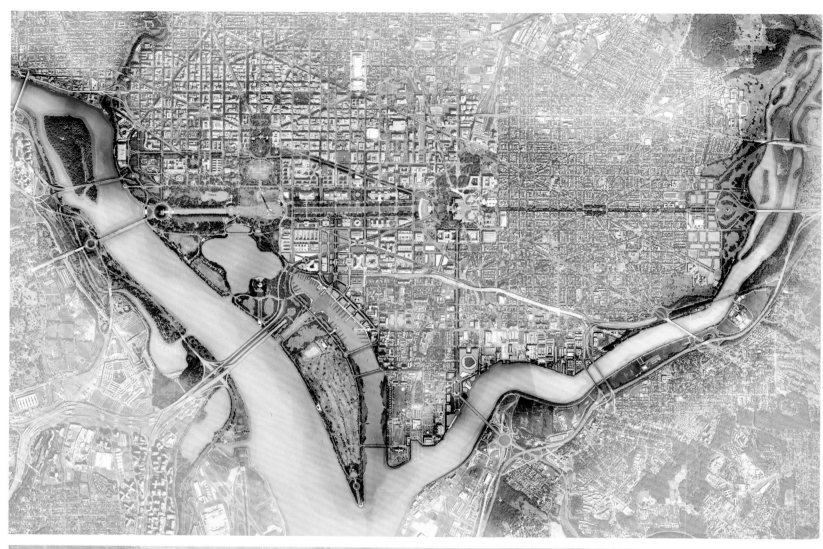

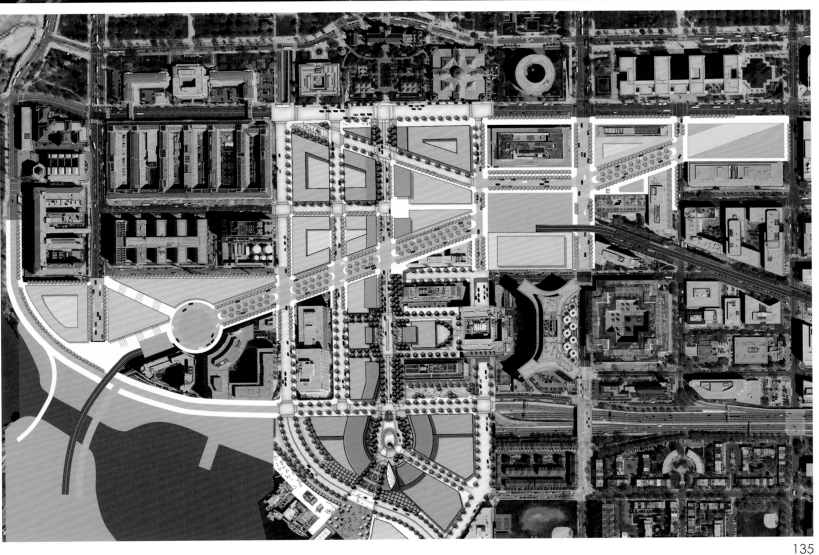

135

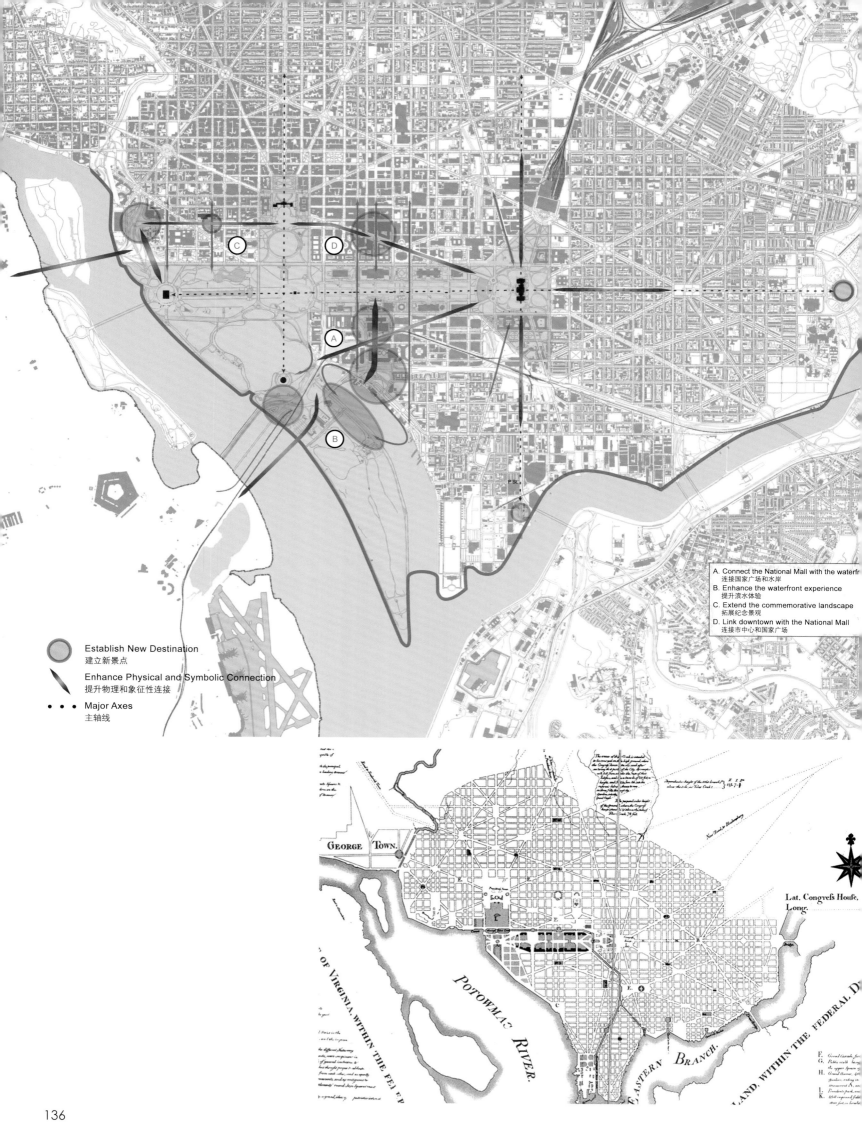

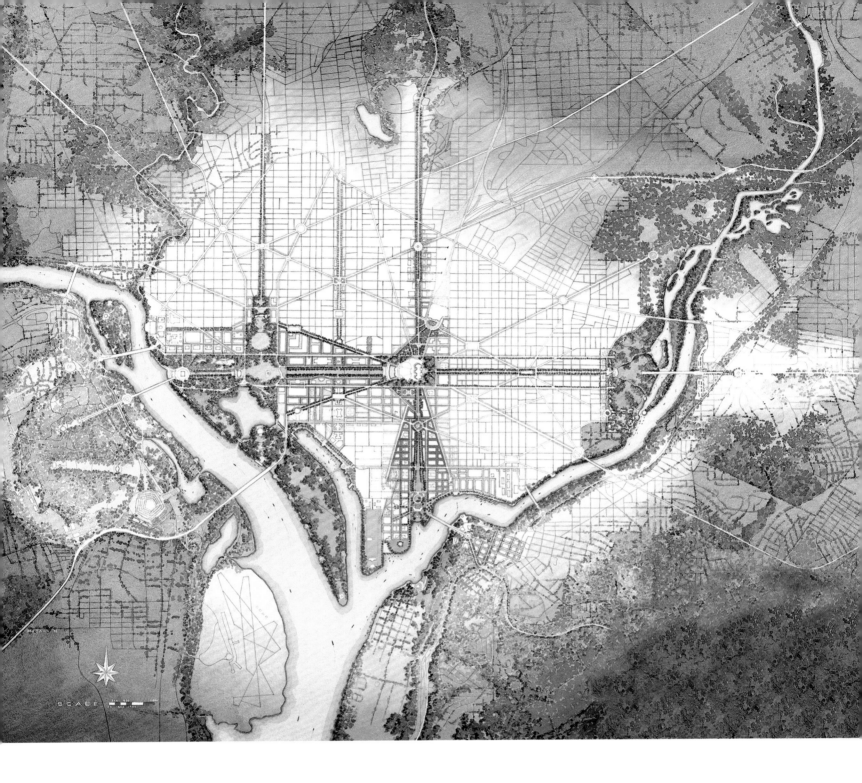

项目特色

纪念碑核心区框架规划由美国国家美术委员会和美国首都规划委员会两个政府部门发起，旨在将国家广场周边的联邦区域改造成为充满活力的景点并且提升城市、国家广场与滨水区之间的联系。框架规划从首都的前景规划环境入手，是指导未来30年决策和投资的实用工具，进一步完善了华盛顿现有的城市规划。

规划包含一系列用于保护国家广场的策略，为文化设施和纪念工作提供独具特色的背景布置，克服了国家广场与其周边城区之间的障碍，同时也提升了华盛顿纪念碑核心区作为标志性可持续区域的形象。框架规划的最初意图是将国家纪念场地纳入纪念碑核心区之中。此外，在保留了市中心的联邦办公地点的同时，它还为新公园、基础设施、交通改革以及多功能开发工程提供了机会。规划由一家基金会引导，以最佳的实践实现了高质量的城市规划、睿智型增长和建筑、基础设施以及城市生态环境的可持续策略。

Ryerson University Master Plan
怀尔逊大学总体规划

Jury Comments:
With its thoughtful connection to the area transportation system and extensive integration with the city, this plan is a decidedly 21st Century response to co-development, including funding and potential integrations of uses within a tight timeframe.

评委评语：
项目巧妙地与当地交通系统以及城市网络连接在一起，是21世纪大学在合作开发（包括在有限时间内的资金投入和潜在使用价值）的典范。

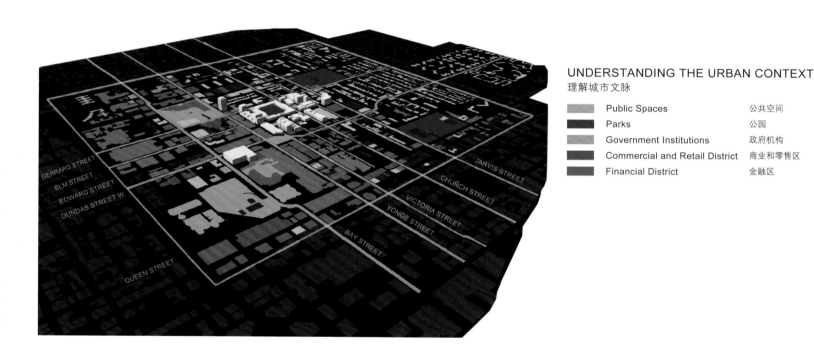

UNDERSTANDING THE URBAN CONTEXT
理解城市文脉

- Public Spaces — 公共空间
- Parks — 公园
- Government Institutions — 政府机构
- Commercial and Retail District — 商业和零售区
- Financial District — 金融区

Notes of Interest

While the Master Plan was developed to deal effectively with the Ryerson University (RU) campus' deficiencies, it ultimately foregrounds Ryerson as a city building, and a model for the 21st Century urban university. Each goal of the Master Plan is defined by a series of principles, and together, they form the flexible framework which will guide the growth of Ryerson University. These goals are: urban intensification, people first (pedestrianization of the urban environment), and a commitment to design excellence.

The Master Plan represents the collective input of the entire University community over an 18-month development period. The consortium architects/urban planners/financial advisors worked closely with Ryerson University senior administration, faculty, students, and the RU Board to articulate the university in clear, concise and implementable terms. Unlike conventional master plans that approach a site and provide specific guidelines in terms of use and density, massing, etc., the RU Master Plan uses demonstration sites, a building site, street, or open space, to demonstrate the application of goals and guidelines. Given the demands of an evolving academic program, the economic imperative to look at "co-development" with commercial or residential development partners, the demonstration sites are presented in the Master Plan document as tools to suggest design directions consistent with the principles.

Consultant: Curran McCabe Ravindran Ross Inc., Gottschalk+Ash International
Engineer: Stantec Consulting, Halcrow Yolles, Crossey Engineering
Owner: Ryerson University

顾问： CMRR公司、哥特沙克+艾什国际
工程师： 斯坦泰克咨询公司、合乐·尤莱斯、克罗塞工程公司
所有人： 怀尔逊大学

Architect / 建筑师
Kuwabara Payne McKenna Blumberg Architects and Daoust Lestage, Inc. in association with Greenberg Consultants, Inc. and IBI Group
KPMB建筑事务所和道斯特·莱斯塔奇公司与格林伯格咨询公司和IBI集团合作

Location / 项目地点
Toronto, Ontario
安大略省，多伦多

Photo Credit / 图片版权
© Kuwabara Payne McKenna Blumberg Architects,
© Daoust Lestage Inc.,
© Greenberg Consultants Inc., © IBI Group
KPMB建筑事务所、道斯特·莱斯塔奇公司、格林伯格咨询公司、IBI集团

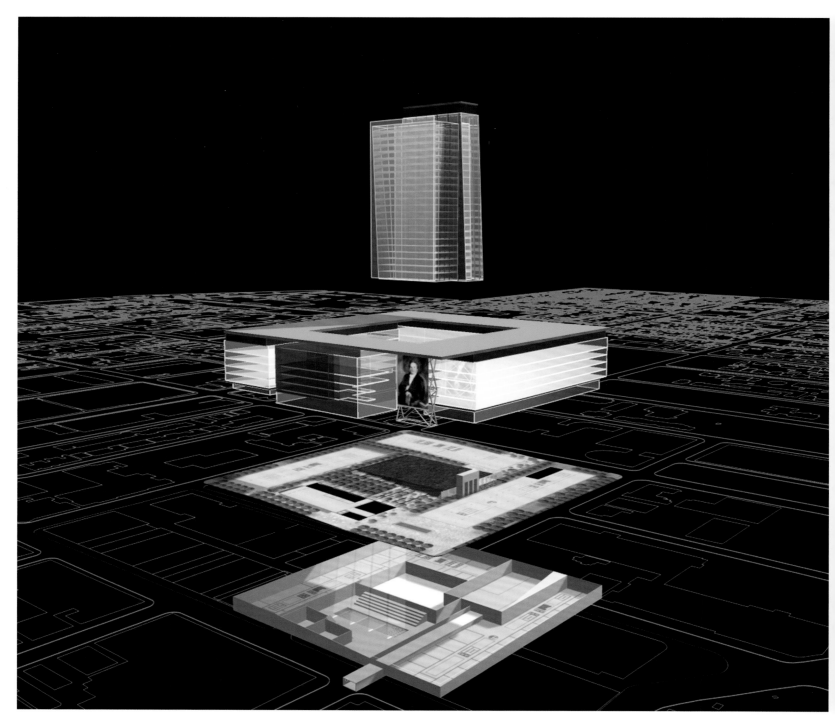

项目特色

总体规划旨在弥补怀尔逊大学校园设计中的不足，最终让怀尔逊前楼改造成为一座城市建筑，作为21世纪城市大学的典范。总体规划的每个目标都配有一系列指导原则，它们共同组成了指导怀尔逊大学发展的灵活框架。这些目标是：城市集约化、以人为本（城市环境中的人行道）和优秀设计方案。

总体规划在18个月的开发期内展现了整个大学社区的整体合作方案。建筑师、城市规划师、财政顾问与怀尔逊大学高级行政部门、教职员工、学生以及董事会通力合作让规划与学校清晰、明确而可行性学期课程紧密结合。与传统规划不同，项目并没有为场地提供详细的规划（功能、密度、体量等），而是利用示范场地（建筑场地、街道或者开放空间）来展示项目目标和指导原则。出于不断演变的学术项目需求，与商业和住宅开发商的"合作开发"将为项目提供经济基础。总体规划中的示范场地将展现设计的指导方针。

Savannah East Riverfront Extension
萨凡纳东部河岸区域扩建

Jury Comments:
This very sensitive addition to one of America's historic treasures is simply the right thing to do and is carried out with a real understanding, pride, and careful analysis.

评委评语：
这个针对美国历史宝藏区域之一的扩建项目十分必要，它经过精心的规划和分析终于得以实施。

COMPATIBLE BUILDINGS

Architectural standards should follow to the extent possible the Manual for Development in Savannah's Historic District.
BUILD-TO LINES AND ENTRANCES: Continuity of building frontage produces a sense of defined space in the public realm. Build-To Lines are established within each block designating a range for required number of minimum entrances in each block. Blocks intended for retail uses have the most permeable building fronts with minimum access occurring every 50 feet, those intended for residential or large commercial uses provide minimum access every 50 to 100 feet.
HEIGHT AND MASS: Consistency of building heights and mass create visual continuity in the streetscape and the skyline. Maximum heights measured in stories have been established in each block, ranging from 4 to 10 stories. A parking level is counted as a story. Building façades over 60 feet in width should be broken down vertically to reduce their mass and create a human scale. Horizontal articulation of buildings should include an identifiable base, body and cap.
COLOR AND TEXTURE: Building walls should be of traditional masonry material such as brick or true stucco. Surfaces should be detailed to provide visual texture and human scale. On taller buildings fronting the water, the colors of surfaces above 4 stories shall be darker and visually recessive in order to not dominate the skyline from the river.
MIXED USES & SUBDIVISION OF BLOCKS: Blocks should be subdivided to allow for a diverse range of building types and lot sizes. Primary uses may be mixed within blocks and within buildings. Fee-simple lots may be subdivided within blocks in varied configurations. Subdivided blocks shall allow for each lot to have a minimum of 20 feet of frontage on a primary street or a lane and a minimum lot size of 1,000 square feet.

可兼容建筑

建筑标准应该依照《萨凡纳历史城区开发手册》的规定。
建成线条和入口：建筑正面的连续性在公共领域中打造了一种明确的空间感。每个街区内的建成线条都保证了最少数量的街区入口。零售街区的入口最多，至少每隔15米就有一个；而住宅或大型商业街区的入口间隔为15到30米。
高度和规模：建筑高度和体量的一致性在街景和空中轮廓中打造了视觉连续性。每个街区都限制了楼高，从4层到10层不一。停车场也算作一层。超过18米宽的建筑外立面应该被垂直劈开，以减少它的体量，打造更人性化的尺度。建筑之间的水平连接包括具有形象感的底座、楼体和顶部。
色彩和纹理：建筑墙面应该采用传统的砖石结构，例如砖块或灰泥。墙壁表面的细部设计应该提供视觉质感和人性尺度。在较高的滨水建筑中，4层以上的外立面的色彩应该更深，不应比河流更加出彩。
混合用途&街区分划：街区应该被进一步分划，以保证建筑类型和场地尺寸的多样化。主要的功能价值被混合在街区和建筑之内。简单的场地在街区中以不同的配置进行分划。被分划的街区要让每块场地都至少有6米长的临街面，而最小的场地面积为93平方米。

CIVIC MASTER PLAN
城市总平面图

Historic view of the East Riverfront
东部河岸历史照片

Notes of Interest

Savannah's Civic Master Plan for the East Riverfront has two goals, one vision: Public Framework and Private Freedom.

The City of Savannah, Georgia identified a series of large vacant parcels along the boundary of its historic City Center for eastward downtown expansion. The first goal was to successfully grow the historic city plan of Savannah after over 150 years. The second goal was a physical and regulatory framework that would allow the expansion district to evolve into a thriving and authentic urban extension.

The Civic Master Plan for the East Riverfront Expansion was implemented by the City in 2006. It defines 54 acres located to the immediate east of Savannah's National Landmark Historic District along the Savannah River. New city blocks, parks, public

Associate Architect: Niles Bolton Associates
Engineer: Thomas & Hutton Engineering Co.
Landscape Architect:
Michael Van Valkenburgh Associates, Inc.
Sasaki, Reed | Hilderbrand
Owner: City of Savannah, ALR Ogelthorpe LLC

合作建筑师：奈尔斯·波顿事务所
工程师：托马斯&休顿工程公司
景观建筑师：迈克尔·凡·瓦肯伯格事务所、佐佐木事务所里德|希尔德布兰德
所有人：萨凡纳市政府、ALR奥格尔索普公司

Architect / 建筑师	Location / 项目地点	Photo Credit / 图片版权
Sottile & Sottile	Savannah	© Sottile & Sottile
苏蒂尔&苏蒂尔	萨凡纳	苏蒂尔&苏蒂尔

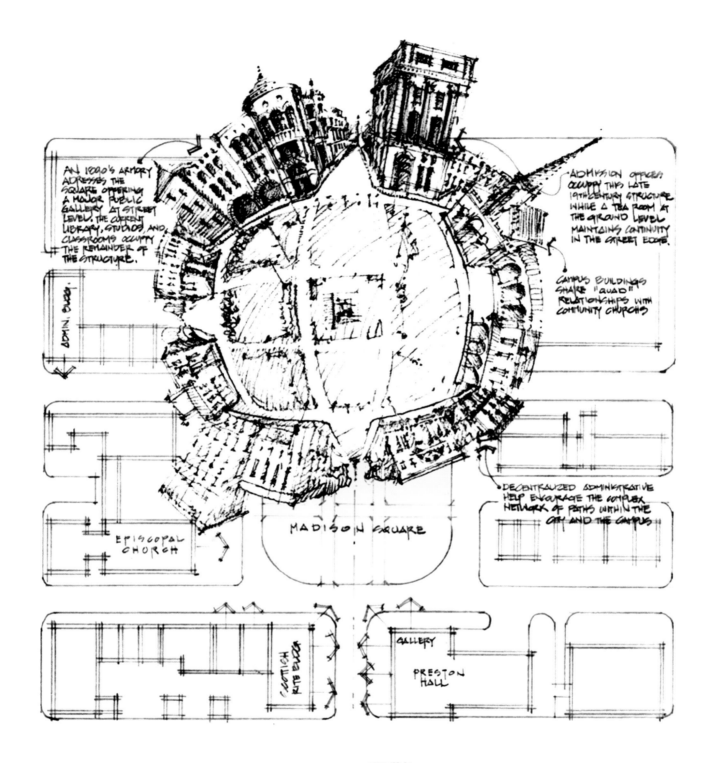

项目特色

萨凡纳东部河岸区市政总体规划有两个目标：公共框架和个人自由。
萨凡纳市沿着它的历史市中心边界划分出一系列大型空地作为城市向东扩张用地。市政府的首要目标是让拥有150余年历史的萨凡纳城市规划得到发展。第二个目标是打造一个能够让扩张区域演变成兴盛的城区的管理框架。
萨凡纳东部河岸区市政总体规划于2006年得以实施。它对萨凡纳国家地标历史区以东沿着萨凡纳河的54英亩土地进行了规划。全新的城市街区、公园、公共空间和河畔走道正在建设之中。项目的私营部分将在未来10年逐步实施，预计成本为8亿美元。
市政总体规划的重要意义并不体现在现有的开发工程，而是体现在未来其所打造的公共区域。设计过程延续了五年之久，针对城市、市民、业主和开发资金等方面召开了各种专家研讨会。最终形成了一个统一的城市可持续发展规划日程，符合城市扩张的基础结果。

spaces and a 2,000-foot river walk extension are currently under construction. The initial private sector build out is expected in 10 years at an estimated cost of 800 million dollars.
However, the significance of the Civic Master Plan is not the magnitude of the current development effort, but in the longevity of the public realm that is created. The design process evolved over a five-year timeframe including multiple public charrettes between the City, citizens, property owners and development interests. The outcome of the process was a unified agenda of sustainable urban growth and the creation of a Civic Master Plan as the fundamental mechanism for urban expansion.

BLOCK SUBDIVISION
街区分划举例

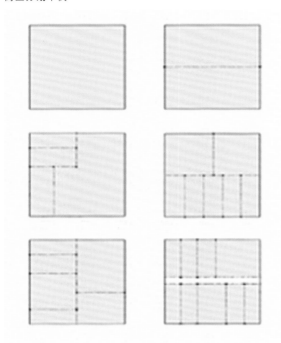

BLOCK PLAN: The single most important defining element of the Civic Master Plan is the street and block plan, connecting this large open property to the historic patterns of the City of Savannah. The street and block plan organizes the site, provides public access to the waterfront, respects view corridors, improves circulation, and creates small blocks to accommodate a range of uses and open spaces. It is the basis for the official mapping of streets, parks, and other public spaces that will shape the future of the public realm.

STREETS AND PARKING: A network of interconnected streets provides circulation throughout the area. Street sections are compact with 10 foot travel lanes on most two-way streets and 12 foot travel lanes on one-way streets. Curb radiuses are minimized to between 6 and 12 feet. Parallel parking is provided on all streets on either one or two sides. Parallel parking is accommodated in 8x20 foot bays. Additional off-street parking may be provided in surface lots with appropriate landscaping or in structured decks. Off-street parking may not front a build-to line.

SIDEWALKS AND STREET TREES: A network of continuous sidewalks on all streets promotes connections and pedestrian-oriented development. Sidewalks are generally 5 to 6 feet in width, ranging up to 20 feet along retail frontage. Street trees are provided in tree lawns are 6 to 13 feet in width and are located between the sidewalk and the street. Regularly spaced and aligned street trees provide human scale, visual continuity, shade for pedestrians, and a barrier between moving traffic. Trees should be Live Oaks at 40 to 50 foot intervals, or other species compatible to those found in the city center.

PUBLIC SPACES: A variety of parks are created with an emphasis on the riverfront. The extension of the river walk is an integral element in connecting these spaces. Parks should be designed to accommodate a variety of passive uses. Multiples focal points should be developed which may include a playground, amphitheater, fountains, or connection to water travel.

街区规划：市政总体规划中最重要的决定元素就是街道和街区规划，它们使这个大型开放式项目与萨凡纳城的历史网络连接在一起。街道和街区规划组织了场地，提供了通往河畔的公共通道，尊重了视野长廊，提升了交通路线并且打造了具有各种用途和开放空间的小型街区。它是官方绘制街道、公园和其他公共空间的地图的基础，将会塑造公共领域的未来。

街道和停车场：相互连通的街道网络为整个区域提供了交通路线。大多数双向街道的截面紧密结合了3米宽的旅游小巷，而单向街道则与3.6米宽的小巷相连。路边半径最缩小到1.8米到3.6米。所有街道都在一侧或两侧平行停车位，大小为2.4米x3.6米。非临街停车场将被设在地表，配有适当的景观设计。非临街停车场可能不会正式建成线路。

人行道和行道树：街道两侧连续的人行道促进了连接和步行开发项目。人行道的宽度普遍在1.5米到1.8米之间，在零售街面前可以拓展到6米宽。行道树的区域宽度为1.8米到4米之间，设在人行道和街道之间。规律排列的行道树具有人性尺度、视觉连续性，为行人提供了树荫并在人与机动车之间建立屏障。树木为橡树或其他市中心常见的树种，间距在12米到15米之间。

公共空间：一系列公园突出了水岸的特色。河畔走道的扩展连接了各个空间。公园的设计拥有一系列被动价值。项目将包含多重焦点：操场、露天剧场、喷泉与水路的连接。

Savannah's East Riverfront Extension adjacent to the Historic City Center
萨凡纳东部河岸区域扩建，紧邻老城中心

Realization of the Plan extension, 2008
实现规划扩张，2008年

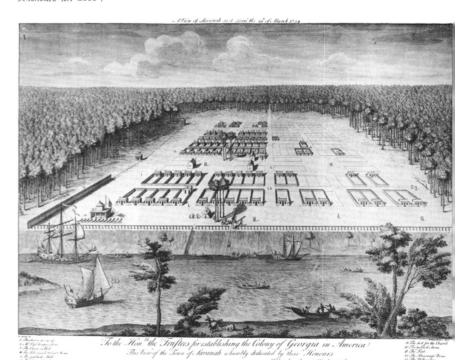

1733年

Revisiting the Earliest Ownership Patterns (Above)
重回最早的所有权模式（上图）

Savannah 1798, Highlighting East Riverfront Wards
萨凡纳1798年，突出东河岸三角区

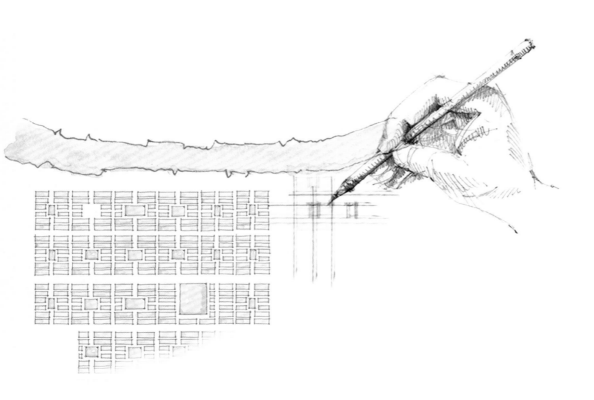

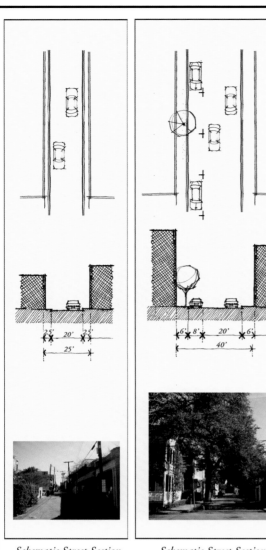
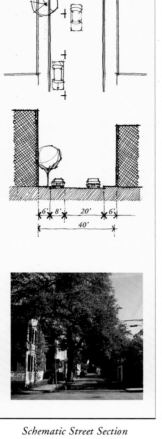
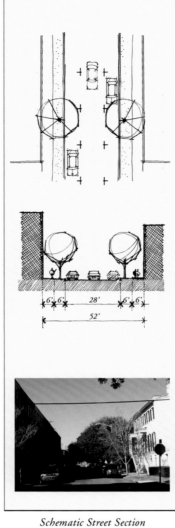
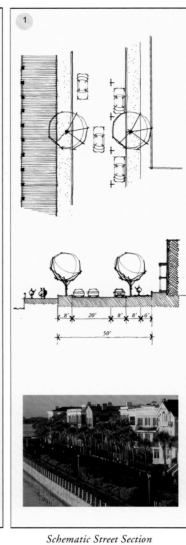
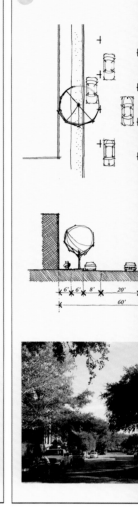

Schematic Street Section	*Schematic Street Section*	*Schematic Street Section*	*Schematic Street Section*	*Schematic Street*
25'LANE	**40'STREET**	**52'STREET**	**50'RIVERFRONT STREET**	**60'STRE**
25' 小巷	40' 街道	52' 街道	50' 河岸街道	60' 街道

STREET TYPES
East Riverfront Civic Master Plan
SAVANNAH GEORGIA

街道类型

东部河岸的街道和公共空间

乔治亚州萨凡纳

EXTENDING CITY STREETS (1 2 3)

Streets, Sidewalks & Street Trees

A network of interconnected streets provides circulation throughout the area. Street sections are COMPACT with 10 foot travel lanes on most two-way streets. Curb radiuses are minimized. Parallel parking is provided on all streets on either one or two sides.

Continuous sidewalks on all streets promote CONNECTIONS and pedestrian-oriented development. Street trees are provided in tree lawns along all primary and secondary street. Regularly spaced and aligned street trees provide HUMAN SCALE, visual continuity, shade for pedestrians and safety from moving vehicles.

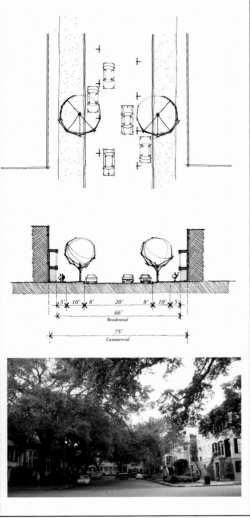
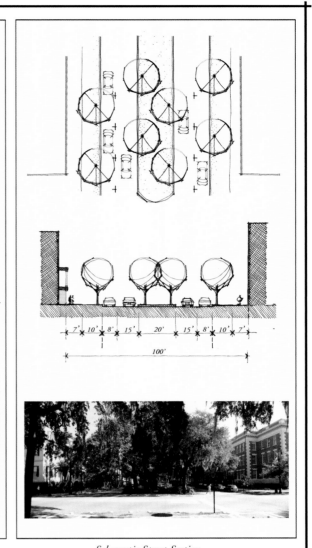

Schematic Street Section

66'/75' STREET
66'/75' 街道

Schematic Street Section

80' STREET
80' 街道

Schematic Street Section

100' BOULEVARD
100' 大道

扩展城市街道（ 1　 2　 3 ）

街道、人行道和行道树

连通街道的网络
相互连通的街道网络为整个区域提供了交通路线。大多数双向街道的截面紧密结合了 3 米宽的旅游小巷，而单向街道则与 3.6 米宽的小巷相连。路边半径最缩小到 1.8 到 3.6 米。所有街道都在一侧或两侧平行停车位，大小为 2.4 米 x3.6 米。

街道两侧连续的人行道促进了连接和步行开发项目。人行道的宽度普遍在 1.5 米到 1.8 米之间，在零售街面前可以拓展到 6 米宽。行道树的区域宽度为 1.8 米到 4 米之间，设在人行道和街道之间。规律排列的行道树具有人性尺度、视觉连续性，为行人提供了树荫并保证了机动车的安全性。

City of Savannah
Metropolitan Planning Commission
Savannah Development & Renewal Authority

In collaboration with

Savannah River Landing, LLC
The Ambling Companies

Thomas & Hutton Engineering Co.
Sottile & Sottile *Urban Design*

February 24, 2006

Information contained herein has been compiled from various sources. Due to availability of data, it does not claim complete accuracy. It is intended instead to provide an overview and analysis of urban conditions.

© Copyright 2006, Sottile & Sottile

The U.S. House Office Buildings Facilities Plan and Preliminary South Capitol Area Plan
美国众议院办公设施规划和南部国会区初步规划

Jury Comments:
From the three adjacent House office buildings south to the Anacostia River, this plan pays a great deal of respect to the original L'Enfant plan, shifts space use to maximize adjacency priorities, and pays considerable attention to facility restoration and the incorporation of sustainable-design elements such as runoff control...

评委评语：
从三座相连的议会办公楼到安娜考斯蒂亚河，该项目充分尊重初始规划设计，变换空间用以最大化四周空间，并注重设施的修复和可持续设计元素（例如径流控制）的运用。

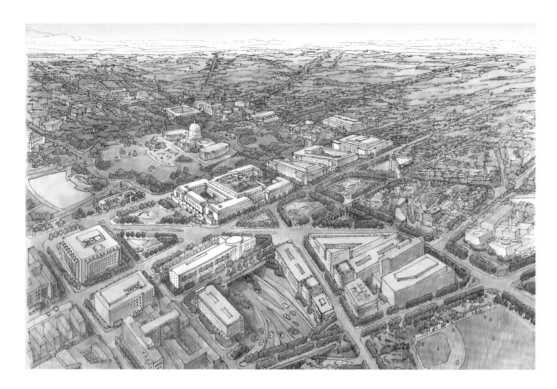

Notes of Interest

The U.S. Capitol Complex in Washington, D.C. is one of the most significant and sensitive places in our country. Within it, the U.S. House of Representatives is its largest component. The House Office Buildings Plan and South Capitol Area Plan defines a vision for fulfilling the current and future space and functional needs of the House, serves as the basis for organizing, budgeting, and funding its long-range capital improvements, and establishes an interface with the future re-development of the South Capitol District from the U.S. Capitol Complex to the Anacostia River.

The project area comprises the 177-acre South Capitol District, and focuses on land within it that is part of the U.S. Capitol Complex, containing some 2.56 million square feet and 5,772 parking spaces assigned to the House, whose needs were projected to grow for 2025 to 3.16 million square feet and 7,283 spaces. The goals, as outlined by Congress, included accommodating growth, improving security, improving transit links, preserving historic assets, upgrading open spaces, complementing new urban development south of the Complex, and developing an overall sustainability framework for the district. The plan also looks beyond 2025 to 2050, to the vision of the National Capitol Planning Commission's Legacy Plan, anticipating the removal of a railroad viaduct and the I-295/395 superstructure. The extension of the Canal Park would complete the integration of the South Capitol District with the U.S. Capitol Complex, to finally heal a scarred urban fabric.

Consultant: Carter Goble Associates, Inc., Booz Allen Hamilton Inc.
Engineer: Louis Berger Group, Inc.
Landscape Architect: Wallace Roberts & Todd
Owner: Architect of the Capitol
顾问： 卡特·戈布尔事务所、博思艾伦咨询公司
工程师： 路易斯·伯吉集团
景观建筑师： 华莱士·罗伯兹&托德公司
所有人： 美国国会大厦建筑师

Architect / 建筑师
Wallace Roberts & Todd, LLC
华莱士·罗伯兹&托德公司

Location / 项目地点
Washington, D.C.
华盛顿

Photo Credit / 图片版权
© Architecture Research Office
建筑研究办公室

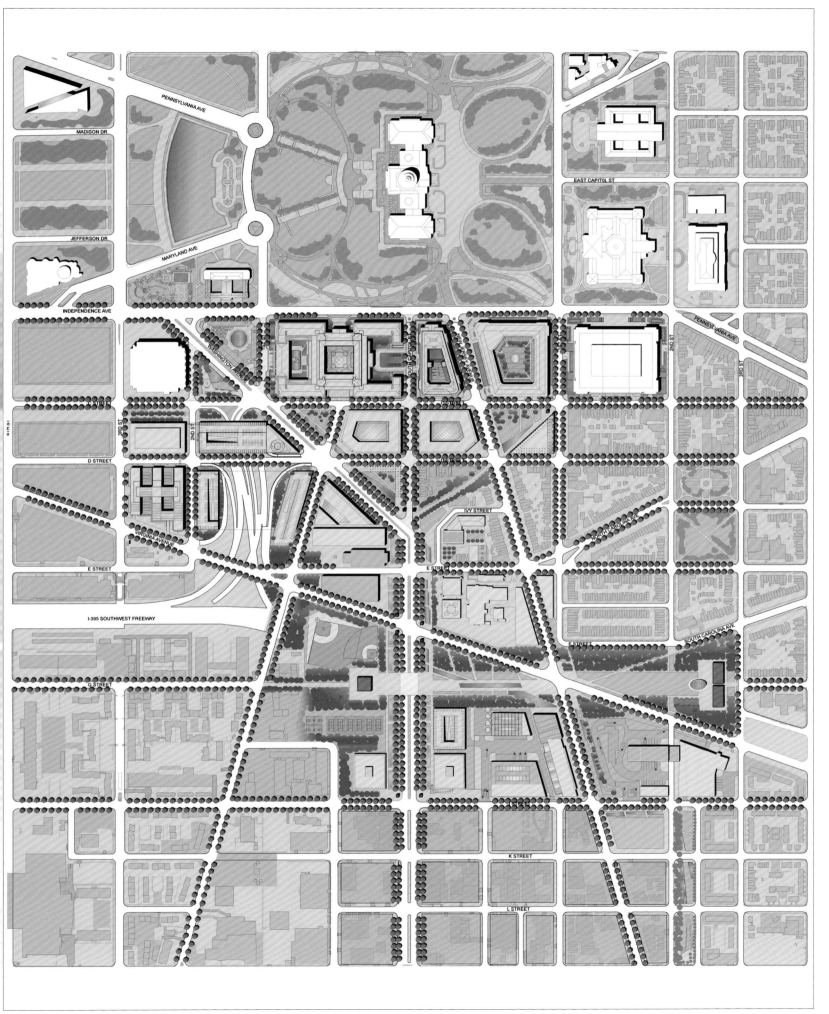

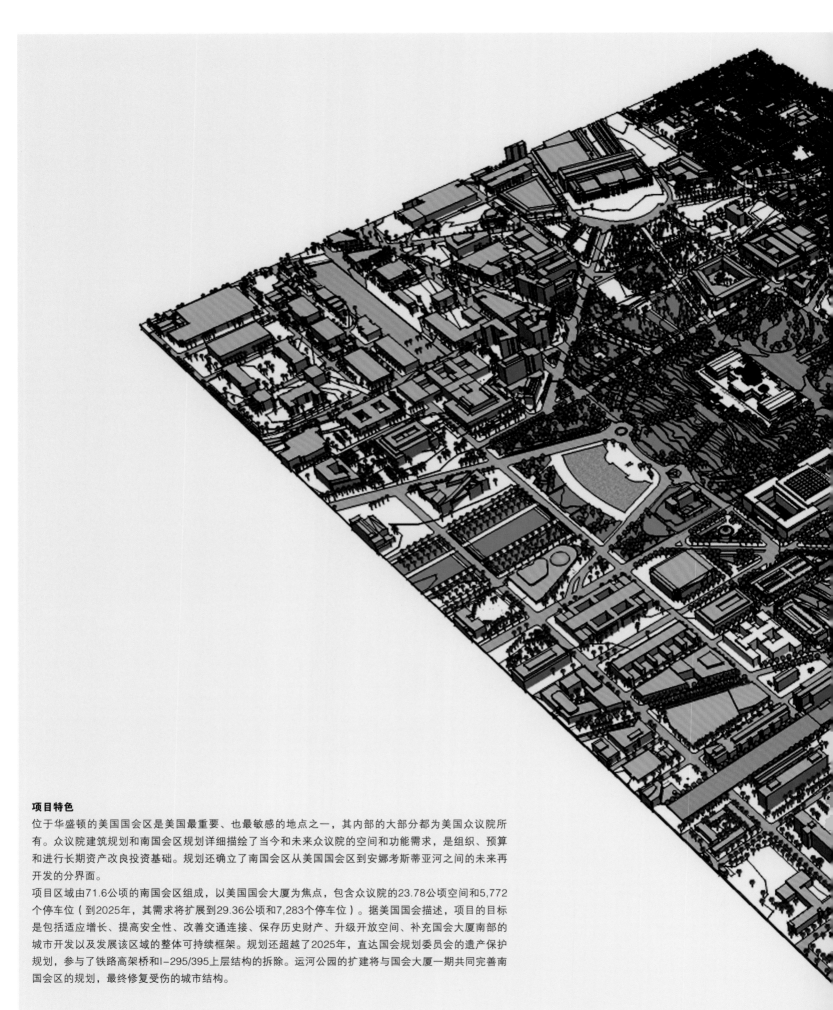

项目特色

位于华盛顿的美国国会区是美国最重要、也最敏感的地点之一，其内部的大部分都为美国众议院所有。众议院建筑规划和南国会区规划详细描绘了当今和未来众议院的空间和功能需求，是组织、预算和进行长期资产改良投资基础。规划还确立了南国会区从美国国会到安娜考斯蒂亚河之间的未来再开发的分界面。

项目区域由71.6公顷的南国会区组成，以美国国会大厦为焦点，包含众议院的23.78公顷空间和5,772个停车位（到2025年，其需求将扩展到29.36公顷和7,283个停车位）。据美国国会描述，项目的目标是包括适应增长、提高安全性、改善交通连接、保存历史财产、升级开放空间、补充国会大厦南部的城市开发以及发展该区域的整体可持续框架。规划还超越了2025年，直达国会规划委员会的遗产保护规划，参与了铁路高架桥和I-295/395上层结构的拆除。运河公园的扩建将与国会大厦一期共同完善南国会区的规划，最终修复受伤的城市结构。

King Abdul Aziz International Airport – Hajj Terminal
阿卜杜勒阿齐兹国王国际机场朝觐航站楼

Jury Comments:
The architects created a highly sustainable project well ahead of the green movement; they learned from the way people have inhabited the desert since early civilization – screening the sun, allowing natural light and ventilation. They did so much with so little – few materials, a regular rhythm of structural bays, a simple fabric structure that works as shelter, as environmental control and as a tie to tradition.
The great roof still works as originally designed as a plaza for the pilgrimage. The building is highly regarded for what it offers spatially, spiritually, symbolically, culturally – it has acquired landmark status as an airport and in the region.

评委评语：
建筑师走在绿色运动之前，打造了一个具有高度可持续的项目；他们学习了早期文明的人们在沙漠中生活的方式——遮阳、保证自然采光和通风。他们在有限的资源中创造了辉煌的成就——较少的材料、常规建筑结构、简单的构造结构共同形成了符合传统而又具有环境价值的庇护所。
巨大的屋顶设计成朝圣之行的广场。建筑在空间、精神、象征、文化意义上的价值得到了高度赞赏——它已经成为了当地的地标性建筑。

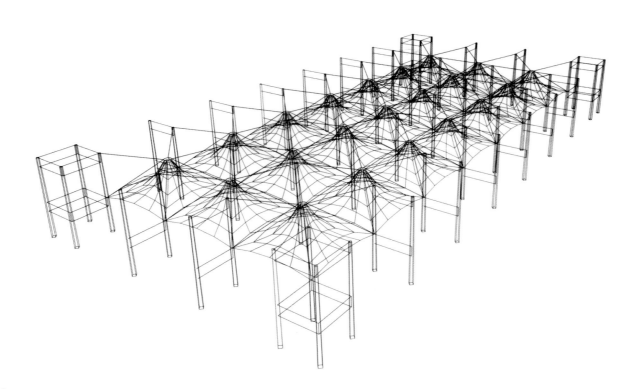

Notes of Interest
The King Abdul Aziz International Airport – Hajj Terminal receives millions of Hajj pilgrims on their way to the Islamic holy site of Mecca through this tented, open-air terminal each year. It was designed and serves as a gathering place of religious fellowship, an improvised campsite for pilgrims waiting to begin their journey, and a point of departure and gateway to Islam's most revered places.
The ritual journey of Muslims to Mecca is one of Islam's five pillars of faith. Pilgrims from around the world travel to Mecca and perform a series of religious rituals. Over a six week span, millions of Muslims undertake this journey. When SOM signed onto the project, the increasing number of Hajj pilgrims had overwhelmed the original Jeddah airport 43 miles west of Mecca, and the firm's Chicago office was commissioned to design a dedicated Hajj terminal there that would only be used during these religious ceremonies. Completed in 1981, the terminal covers 120 acres and 2.8 million square feet.

项目特色
阿卜杜勒阿齐兹国王国际机场朝觐航站楼每年接待上百万的前往伊斯兰圣地麦加的朝圣者。它被设计成一个宗教伙伴聚会的场所，是朝圣者的临时出发营地，也是前往伊斯兰圣地的出发地和门户。
穆斯林前往麦加的宗教之旅是伊斯兰教的五大信仰支柱之一。来自世界各地的朝圣者前往麦加并进行一系列的宗教仪式。在六周时间里，上百万的穆斯林经历了这次旅程。当SOM建筑事务所签下这个项目时，人数激增的朝圣者已经令距离麦加43英里的原吉达机场不堪重负，SOM的芝加哥分公司受委托设计一个只在宗教仪式期间开放的朝觐航站楼。航站楼建成于1981年，占地面积48.6公顷，总建筑面积28公顷。

General Contractor: Hochtief A.G.
Owner: Kingdom of Saudi Arabia Ministry of Defense & Avia
总承包商： 霍克蒂夫公司
所有人： 沙特阿拉伯国防航空部

Architect / 建筑师
Skidmore, Owings & Merrill LLP –
New York and Chicago Offices
SOM建筑事务所——纽约和芝加哥分公司

Location / 项目地点
Jeddah, Saudi Arabia
沙特阿拉伯，吉达

Photo Credit / 图片版权
© Owens-Corning Fiberglas/S. A. Amin/
Skidmore, Owings & Merrill LLP/Prof. em. Herbert Schmidt
欧文斯–康宁·费伯格拉斯/S.A.Amin/、SOM公司/赫伯特·施密特

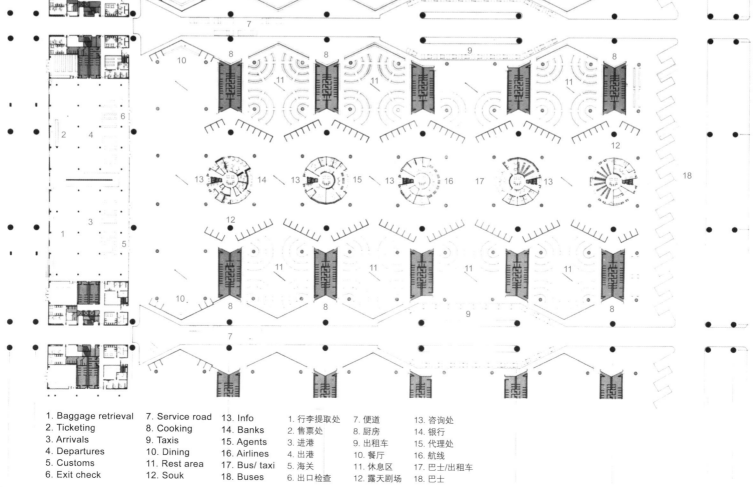

1. Baggage retrieval	7. Service road	13. Info	1. 行李提取处	7. 便道	13. 咨询处
2. Ticketing	8. Cooking	14. Banks	2. 售票处	8. 厨房	14. 银行
3. Arrivals	9. Taxis	15. Agents	3. 进港	9. 出租车	15. 代理处
4. Departures	10. Dining	16. Airlines	4. 出港	10. 餐厅	16. 航线
5. Customs	11. Rest area	17. Bus/ taxi	5. 海关	11. 休息区	17. 巴士/出租车
6. Exit check	12. Souk	18. Buses	6. 出口检查	12. 露天剧场	18. 巴士

2011 INSTITUTE HONOR AWARDS FOR ARCHITECTURE JURY
建筑荣誉奖评委

David E. Miller, FAIA, Chair
The Miller Hull Partnership, LLP
大卫·E·米勒
美国建筑师协会会员；评委会主席
米勒·赫尔建筑事务所

Ashley Clark, Associate AIA
The Littlejohn Group
阿什利·克拉克
美国建筑师协会
小约翰集团

Curtis Fentress, FAIA
Fentress Architects
柯蒂斯·芬特雷斯
美国建筑师协会会员
芬特雷斯建筑事务所

T. Gunny Harboe, FAIA
Harboe Architect, PC
T·加尼·哈尔博
美国建筑师协会会员
哈尔博建筑事务所

David Neuman, FAIA
University of Virginia
大卫·纽曼
美国建筑师协会会员
弗吉尼亚大学

Louis Pounders, FAIA
ANF Architects
路易斯·庞德尔斯
美国建筑师协会会员
ANF 建筑事务所

Sarah Snodgrass
University of Nevada, Las Vegas/
American Institute of Architecture Students Representative
萨拉·斯诺德格拉斯
拉斯维加斯内华达大学
美国建筑师协会学生代表

Allison Williams, FAIA
Perkins & Will
阿里森·威廉姆斯
美国建筑师协会
珀金斯 & 威尔建筑事务所

Jennifer Yoos, AIA
VJAA
珍妮佛·尤斯
美国建筑师协会
VJAA 建筑事务所

2011 Institute Honor Awards
2011年美国建筑师协会建筑 / 室内设计 /
区域和城市规划荣誉奖评委

2011 INSTITUTE HONOR AWARDS FOR INTERIOR ARCHITECTURE JURY
室内设计荣誉奖评委

John Ronan, AIA, Chair
John Ronan Architects
约翰·罗南
美国建筑师协会；评委会主席
约翰·罗南建筑事务所

Jaime Canaves, FAIA
Florida International University,
School of Architecture
杰米·卡纳维斯
美国建筑师协会会员
佛罗里达国际大学建筑学院

Margaret Kittinger, AIA
Beyer Blinder Belle Architects
玛格利特·奇丁格尔
美国建筑师协会
拜尔·布林德·贝勒建筑事务所

Bryan Lewis
The Capital Group Companies
布赖恩·路易斯
美国资本研究集团公司

Brian Malarkey, AIA
Kirksey
布莱恩·马拉奇
美国建筑师协会
奇克斯公司

2011 INSTITUTE HONOR AWARDS FOR REGIONAL AND URBAN DESIGN JURY
区域和城市规划荣誉奖评委

Daniel E. Williams, FAIA, Chair
Daniel Williams Architect
丹尼尔·E·威廉姆斯
美国建筑师协会会员；评委会主席
丹尼尔·威廉姆斯建筑事务所

C.R. George Dove, FAIA
WDG Architecture, PLLC
C·R·乔治·德芙
美国建筑师协会会员
WDG建筑事务所

Vivien Li
Boston Harbor Association
薇薇安·李
波士顿港事务所

Claire Weisz, AIA
Weisz + Yoes Architecture
卡莱尔·维兹
美国建筑师协会
维兹+尤斯建筑事务所

Bernard Zyscovich, FAIA
Zyscovich, Inc.
伯纳德·塞斯科维奇
美国建筑师协会会员
塞斯科维奇公司

David E. Miller, FAIA
2011 Chair,
Institute Honor Awards for Architecture

大卫·E·米勒
美国建筑师协会会员
2011年美国建筑师协会建筑荣誉奖评委会主席

David E. Miller, FAIA, is a founding partner of The Miller | Hull Partnership, a sixty-five person firm in Seattle. Miller|Hull is a fundamentally design oriented firm, emphasizing a rational design approach based on the culture, climate and building traditions of a place. In addition to over 200 awards for design excellence, the firm received the 2003 AIA Architecture Firm Award, which is given to one architectural design practice in the U.S. each year. Three monographs have been published on the firm's work; Ten Houses by Rockport Press, 1999; Miller | Hull, Architects of the Pacific Northwest by Princeton Architectural Press, 2001; and Public Works by Princeton Architectural Press, 2009.

David is an excellent and inspirational designer. In 2006, he received two unique awards: the Washington State University Regents' Distinguished Alumnus Award and the BetterBricks Designer Award, recognizing him as a designer who supports, uses and designs sustainable, high performance, commercial buildings. In 2010 he received the AIA Seattle Chapter Medal, the highest individual award bestowed on an architect by the Chapter.

Currently, David is Chair for the Department of Architecture at the University of Washington, where he is also a tenured professor of architecture. In 2005, David authored, toward a New Regionalism (University Press), which promotes environmental architecture and showcases the work of Northwest architects from Portland to British Columbia. For the 2010 year, he is Co-Chair of the National AIA, Committee on the Environment (COTE), Advisory Group.

大卫·E·米勒（美国建筑师协会会员）是米勒·赫尔事务所创始合伙人。该事务所位于西雅图，共有65名员工，是一家以设计为导向的公司，专注于文化、气候和建筑传统的合理设计。除了200多项优秀设计奖之外，公司还获得了2003年美国建筑师协会公司奖（该奖项旨在奖励美国每年出色的建筑设计公司）。公司曾经出过三本作品集：《十座住宅》（1999；洛克波特出版社）、《米勒·赫尔事务所——太平洋西北海岸的建筑师》（2001；普林斯顿建筑出版社）、《公共作品》（2009；普林斯顿建筑出版社）。

大卫是一位卓越而富有灵感的设计师。2006年，他获得了两个独特的奖项：华盛顿州立大学董事会的杰出校友奖和好砖块设计师奖（该奖项认可了他作为一个设计师在支持、运用和设计可持续、高性能的商业建筑中所作出的贡献）。2010年，他获得了美国建筑师协会西雅图分会奖章——西雅图分会向建筑师颁发的最高独立奖项。

目前，大卫是华盛顿大学建筑部的主席，同时还在那里担任建筑学终身教授。2005年，大卫编写了《朝向新地方主义》（大学出版社），宣传了环境建筑并展现了太平洋西北部（从波特兰到不列颠哥伦比亚）建筑师的作品。2010年，他担任了美国建筑师协会、环境委员会、咨询小组的联席主席。

© David E. Miller 大卫·E·米勒

John Ronan, AIA
2011 Chair,
Institute Honor Awards for Interior Architecture

约翰·罗南
美国建筑师协会
2011美国建筑师协会室内设计荣誉奖评委会主席

© Michelle Litvin 米歇尔·利特文

John Ronan is founding principal of John Ronan Architects in Chicago, founded in 1997. He holds a Master of Architecture degree with distinction from the Harvard University Graduate School of Design and a Bachelor of Science degree with honors from the University of Michigan. In 1999, he was a winner in the Townhouse Revisited Competition staged by the Graham Foundation and his firm was the winner of the prestigious Perth Amboy High School Design Competition in 2004, a two-stage international design competition to design a 472,000 square foot high school in New Jersey. In December 2000, he was named as a member of the Design Vanguard by Architectural Record magazine, and in January 2005 he was selected to The Architectural League of New York's Emerging Voices program. In 2006 he was featured in the Young Chicago exhibition at the Art Institute of Chicago. His work has been exhibited in galleries throughout the U.S., including the Graham Foundation, the Art Institute of Chicago and The Architectural League of New York, and his work has been featured in numerous international publications. A monograph on his work, entitled Explorations, published by Princeton Architectural Press was released in 2010. John is currently an Associate Professor at the Illinois Institute of Technology College of Architecture, where he has taught since 1992.

1997年，约翰·罗南在芝加哥创办了建筑事务所。他拥有哈佛大学设计研究生院建筑学硕士学位和密歇根大学理学学士学位。1999年，他获得了由格雷汉姆基金会所举办的市政厅重游竞赛的优胜。2004年，他的公司获得了著名的珀斯安波易高中设计竞赛。这个两阶段国际设计竞赛要求在新泽西设计一个占地约4.4公顷的高中。2000年12月，他被《建筑实录》杂志评为设计先锋。2005年1月，他被纽约建筑社团选入新兴之声项目。2006年，他在芝加哥美术馆所举办的"年轻的芝加哥"展览进行了个人展。他的作品在美国各大美术馆中进行了展览，其中包括格兰汉姆基金会、芝加哥美术馆和纽约建筑联盟，同时也在大量国际出版物中出版。他的学术论文《期望》于2010年被普林斯顿建筑出版社出版。约翰现在是伊利诺伊理工大学建筑学院的副教授，他自1992年起一直在此任教。

Daniel E. Williams, FAIA, APA
2011 Chair,
Institute Honor Awards for Regional & Urban Design
丹尼尔·E·威廉姆斯
美国建筑师协会会员；美国心理协会
2011美国建筑师协会区域和城市规划荣誉奖评委会主席

© Megan H. Williams 梅根·H·威廉姆斯

Daniel E. Williams, is a Fellow in the American Institute of Architects and is an internationally recognized expert in sustainable architecture and urban and regional design. Mr. Williams is a member of the experts team for the Clinton Climate + Initiative, advising on projects in Toronto and London. He served as 2006 chair of the AIA's Sustainability Task Group and sat on the national advisory council for United States Environmental Protection Agency – NACEPT.

He participated in the development, the 2010 Council of Mayor's resolution that will reduce carbon emissions by 50%; presented Watershed Planning Initiatives at the Center for Neighborhood Technologies in Chicago; wrote and chaired the AIA/EPA grant Water + Design; co-wrote the Barcelona Declaration on Sustainability; and has worked with dozens of communities around the country, creating master plans with the residents – specifically to assist in the rebuilding of towns and cities after natural disasters and the associated impacts from climate change.

In 2003 he chaired the National Committee on the Environment for the American Institute of Architects and chaired the Task Force on the Environment and Energy for the Congress for the New Urbanism from 1996 – 2000, and won the first passive design award in Architecture from NASA in 1980.

His work on post-disaster smart growth urban and regional design projects won the 1999 and 2000 National Honor Award for Urban and Regional Design from the American Institute of Architects' and the Catherine Brown Award for Urban Design in the American Landscape in 1999. His projects range in scale from "off the grid" residences to regional master plans of thousands of square miles – these designs integrate issues in ecology, economic development, transportation, agricultural preservation, education, water resource protection, smart growth and climate change. Named Eminent Scholar and Distinguished Alumni in 2000 at the University of Florida, his book Sustainable Design: Ecology, Architecture and Planning was published Earth Day 2007 by John Wiley & Sons. He is presently working on a book titled Design with Climate-Change.

丹尼尔·E·威廉姆斯（美国建筑师协会会员）是可持续建筑及区域和城市规划方面的国际知名专家，也是克林顿气候方案在多伦多和伦敦的项目专家团队成员之一。2006年，他曾担任美国建筑师协会可持续任务集团主席，也进入了美国环境保护署的国家顾问委员会。他参与了2010市长决议会议（致力减少50%的碳排放量）的开发；在芝加哥社区技术中心提出了分水岭规划方案；撰写并领导了美国建筑师协会/环境保护署拨款的水+设计项目；合作撰写了巴塞罗那可持续发展宣言；并且与数十个美国国内社区及其居民共同打造了住宅总体规划——特别是帮助重建了许多因遭受气候变化而引起的自然灾害破坏的城镇。

2003年，威廉姆斯担任了美国建筑师协会国家环境委员会主席。1996年-2000年间，他担任了新城市主义环境与能源特别小组的主席，并且于1980年获得了美国国家宇航局颁发的第一个被动式建筑设计大奖。

他的灾后高效增长城市与区域规划项目获得1999年和2000年美国建筑师协会城市与区域规划国家荣誉奖，并获得1999年凯瑟琳·布朗美国景观城市规划大奖。他的项目范围从偏远住宅到绵延数千平方公里的区域整体规划——涉及生态、经济开发、交通、农业保护、教育、水资源保护、高效增长和气候变化等多方面。

2000年，威廉姆斯被佛罗里达大学列为杰出学者和著名校友。他编写的《可持续设计：生态、建筑和规划》一书在2007年世界地球日由约翰·威利集团出版发行。目前他正在撰写名为《随着气候变化进行设计》的书。

AT&T Performing Arts Center Dee and Charles Wyly Theater

AT&T 表演艺术中心迪和查尔斯·威利剧院

Jury Comments:
This building is an expression of a totally new way to investigate the potential of performative experimentation – completely re-choreographed the way in which one experiences a theater.

评审评语：
建筑探索了一条全新的表演中心设计道路——颠覆了人们的剧院体验。

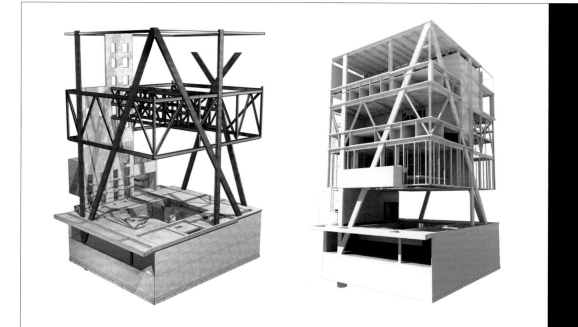

Notes of Interest

The AT&T Performing Arts Center is an 80,300 square-foot, 575-seat "multi-form" theater with the ability to transform between configurations and a performance chamber open to its urban surroundings.

The Dallas Theater Center (DTC)'s previous accommodation – a makeshift metal shed – freed its users from the limitations imposed by a fixed-stage configuration and the need to protect expensive interior finishes.

The Dee and Charles Wyly Theatre overcame these challenges by positioning back-of-house and front-of-house facilities above and beneath the auditorium, instead of encircling it. This unprecedented stacked design transforms the building into a "theater machine" that provides an almost infinite variety of stage-audience configurations and manifests a strong presence in the Dallas Arts District despite its relatively modest size.

Associate Architect: Kendall/Heaton Associates
Engineer: Transsolar Energietechnik, Cosentini Associates, Magnusson Klemencic Associates
Construction Manager: McCarthy Building Companies
Lighting: Tillotson Design Associates
Owner: AT&T Performing Arts Center

合作建筑师：肯德尔/希顿事务所
工程经理：麦卡锡建筑公司
工程师：超日能源技术公司、克森蒂尼事务所、马格努森·克里门斯克事务所
照明设计：狄罗森设计事务所
所有人：AT&T 表演艺术中心

Architect /建筑师
REX | OMA
REX | OMA

Location /项目地点
Dallas
德克萨斯州，达拉斯

Photo Credit /图片版权
© Iwan Baan Photography
伊万·班摄影

项目特色

AT&T表演艺术中心是一个总面积7,460平方米、拥有575个坐席的多功能剧院，剧院拥有变换配置的功能，并且包含一间面向城市开放的表演室。

达拉斯剧院中心的前身——一个临时的金属屋——让使用者们不必局限于固定的舞台配置，也不必保护昂贵的室内装饰。

迪和查尔斯·威利剧院克服了这些挑战，将后台和前台设施设置在剧院礼堂的上下，而不是环绕着舞台。这个史无前例的叠加设计将建筑转变为一个"剧院机器"，提供了无穷无尽的舞台配置可能，以其较小的规模在达拉斯艺术区呈现了强烈的存在感。

Ford Assembly Building
福特装配楼

Jury Comments:
This renovated facility has improved the region by saving an older building and its embodied energy. It is functioning in a fashion of research and design fabrication not dissimilar to Khan's original intentions.

评委评语：
这座翻新设施通过对旧建筑的改造改善了整个区域的环境。建筑的研究和设计风格与卡恩的原始设计出入不大。

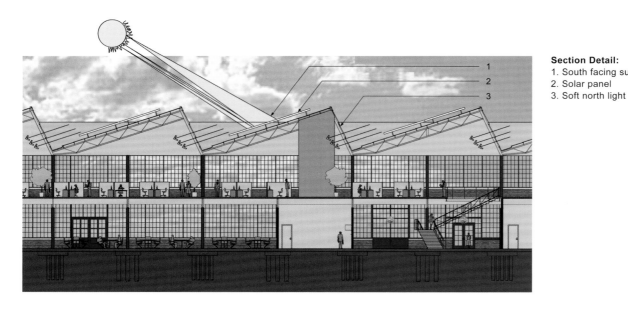

Section Detail:
1. South facing sun
2. Solar panel
3. Soft north light

剖面细部：
1. 南面朝阳
2. 太阳能板
3. 北面柔光

Notes of Interest
The restoration and preservation of the Ford Assembly Building on the San Francisco Bay waterfront saved a historic architectural icon from the wrecking ball and converted a long-vacant auto plant into a current-day model of urban revitalization and sustainability. The 525,000 square foot building had been designed by Albert Kahn for Henry Ford, and constructed in 1931. Following the facility's initial car factory function, the Ford Building had many incarnations, including the famous World War II tank factory "manned" by Rosie-the-Riveters.

Design excellence for the restoration of the Ford Assembly Building meant instilling new purpose and function by revitalizing an already rich historic architectural icon while respecting its existing industrial aesthetic. The restoration allows for the considerable mix of uses at Ford, including offices, manufacturing, R&D labs, support space and retail. Along the east side are loading docks, while the west side has the official front doors of several companies. Favorite amenities for both tenants and visitors include The Craneway event space and The BoilerHouse Restaurant for both tenants and visitors to the Ford campus.

Engineer: Charles M. Salter Associates, Mechanical Design Studio, Inc., The Crosby Group, Gregory P. Luth & Associates
General Contractor: Dalzell Corporation
Historic Preservation: Preservation Architecture
Lighting: Architecture + Light
Owner: Orton Development, Inc.

工程师：查尔斯·M·沙尔特建筑事务所、机械设计工作室、克罗斯比集团、格雷戈里·P·卢斯事务所
总承包商：达尔泽尔公司
历史保护：保护建筑事务所
照明设计：建筑+照明
所有人：奥尔顿开发公司

项目特色
位于旧金山湾的福特装配楼的修复和保护拯救了一座标志性历史建筑，使其免受落锤破碎球的破坏，将一座长期置的汽车工厂改造成为一个城市复兴和可持续设计的典型。这座48,774平方米的建筑由艾伯特·卡恩为福特公司创始人设计，建成于1931年。除了基本的汽车制造厂之外，福特装配楼还曾化身为著名的二战坦克工厂。

福特装配楼的修复设计要求为这座具有丰富历史的建筑注入新的功能，使其重新焕发活力，同时又要尊重建筑的工业美学价值。修复工作为福特公司提供了大量混合空间，包括办公室、生产线、研发实验室、辅助空间和零售区。建筑东侧是装卸码头，西侧是几个公司的正门。租户和访客最喜欢的设施是其中轨道活动空间和锅炉房餐厅。

Architect /建筑师
Marcy Wong Donn Logan Architects
MWDL建筑事务所

Location /项目地点
Richmond, California
加利福尼亚州，里士满

Photo Credit /图片版权
© Billy Hustace
比利·胡斯塔斯

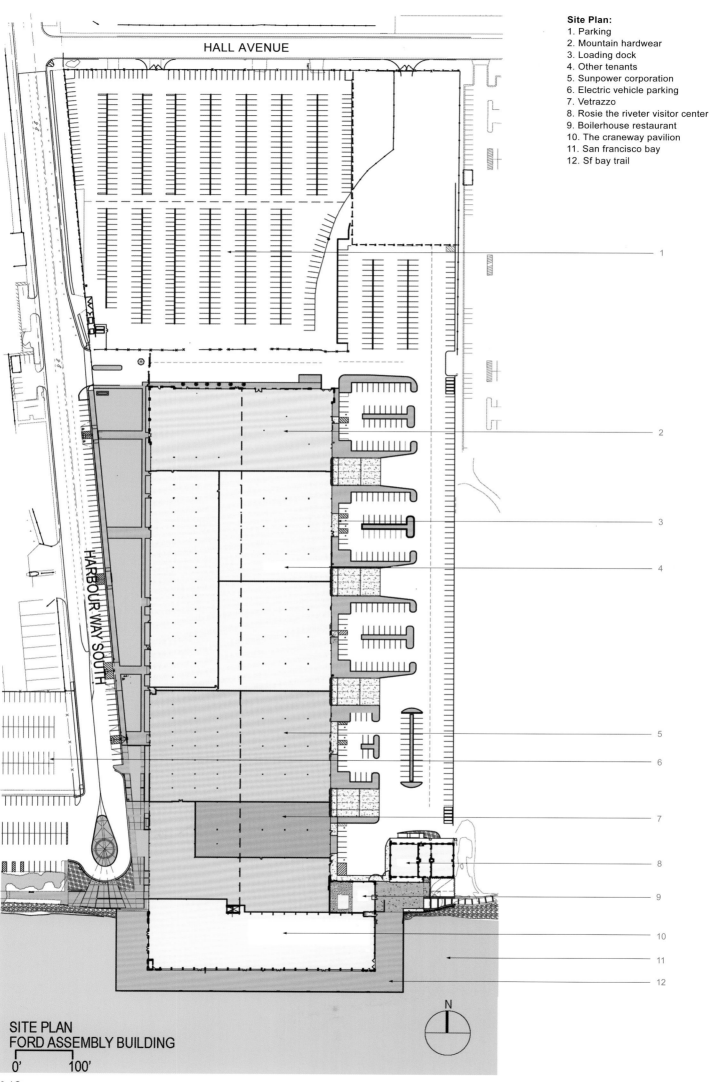

Site Plan:
1. Parking
2. Mountain hardwear
3. Loading dock
4. Other tenants
5. Sunpower corporation
6. Electric vehicle parking
7. Vetrazzo
8. Rosie the riveter visitor center
9. Boilerhouse restaurant
10. The craneway pavilion
11. San francisco bay
12. Sf bay trail

总平面图：
1. 停车场
2. MHW品牌店
3. 装货码头
4. 其他租户
5. 太阳能公司
6. 电动车停车场
7. 维特拉左商店
8. "铆工露丝"游客中心
9. 锅炉房餐厅
10. 起重机轨道亭
11. 旧金山湾
12. 旧金山湾轨道

SITE PLAN
FORD ASSEMBLY BUILDING

Transverse Section Looking North (Below)
北侧横截面（下图）

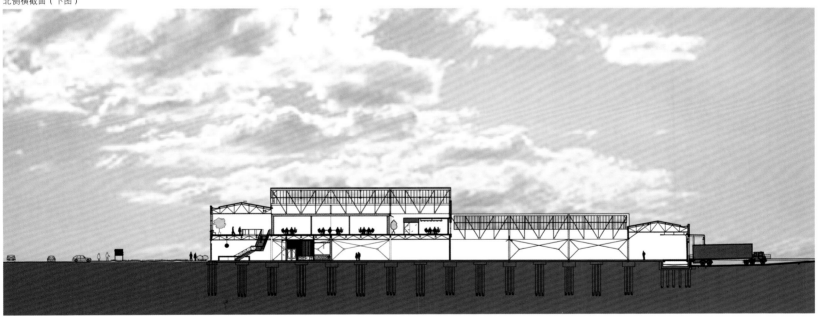

Longitudinal Section Looking West (Below)
西侧纵切面（下图）

Horizontal Skyscraper Vanke Center
水平摩天楼：万科中心

Jury Comments:
This project skips along from mound to mound and manipulates the landscape, while letting it breathe – it builds it up and shapes it into a powerful form above the land with inventive manipulation.
The building is shading the landscape and letting it breathe – integrated sustainability.
A reinvented building type with the building floating over the landscape – dancing on the landscape.

评委评语：
项目沿着高地跳跃，巧妙地处理了景观，让景观随意呼吸。
它独出心裁地脱离地面，形成了有力的造型。
建筑为景观遮阳并让它呼吸，极具可持续性。
这是一个全新的建筑类型——建筑悬浮于景观之上，随着景观而舞蹈。

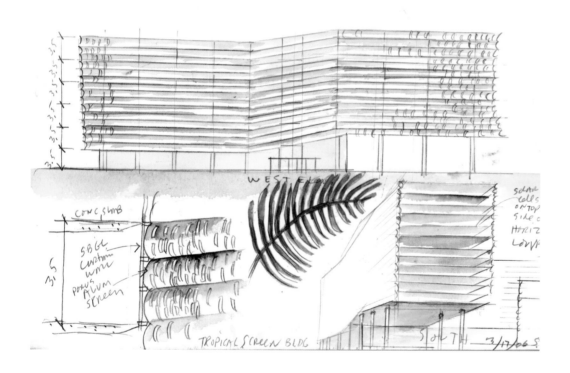

Notes of Interest

The Horizontal Skyscraper Vanke Center stands as a new hybrid model that provides ample public space in a unique way. The building also combines the most forward-thinking sustainable technologies with innovative construction techniques.

The building hovers above the landscape, freeing it for public use and for a unique scheme of ecosystem restoration. Of the 60,000-square-meter site, 28,000 square meters are left unbuilt, and people in the surrounding community have already begun inhabiting the space for leisure. By lifting the building off the ground, the project is both a building and a landscape. The landscape scheme works to minimize run-off, erosion, and other types of environmental damage associated with development.

Additionally, the Horizontal Skyscraper Vanke Center employs some of the most forward-thinking sustainable design strategies. It utilizes greywater recycling, rain water harvesting, green roofs, dynamically controlled operable louvers, and high-performing glass, and a roof of photovoltaic panels that provide 12.5 percent of the total electric energy demand for Vanke Headquarters.

Associate Architect: CCDI
Engineer: Transsolar; CCDI; CABR; CCDI
Lighting: L'Observatoire International
Owner: Shenzhen Vanke Real Estate Co.
合作建筑师：CCDI
工程师：超日公司、CCDI、CABR
照明设计：瞭望国际公司
所有人：深圳万科房产公司

Architect / 建筑师	Location / 项目地点	Photo Credit / 图片版权
Steven Holl Architects 斯蒂文·霍尔建筑事务所	Shenzhen, China 中国，深圳	© Iwan Baan Photography 伊万·班摄影

项目特色

水平摩天楼万科中心是一个全新的混合模型,以其独特的方式提供了充裕的公共空间。建筑结合了最先进的可持续技术与创新建筑技术。

建筑盘旋在景观之上,使下方景观可以用作公共用途,形成了特别的生态保护方案。在60,000平方米的场地面积中,28,000平方米空间未进行建造,周边社区的居民已经开始在这里进行休闲活动。项目通过将结构脱离地面形成了建筑与景观的结合体。景观设计方案能够最小化地面径流、侵蚀以及其他类型的环境破坏。

此外,万科中心采用了最具前瞻性的可持续设计策略。它运用了灰水循环、雨水收集、绿色屋顶、动态控制百叶窗、高性能玻璃以及太阳能光电板屋顶。太阳能光电板将为万科中心提供12.5%所需的能源。

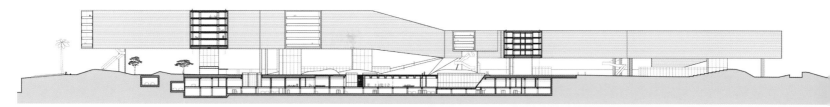

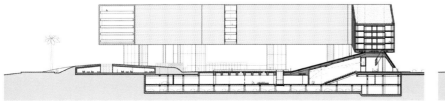
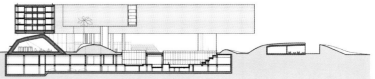

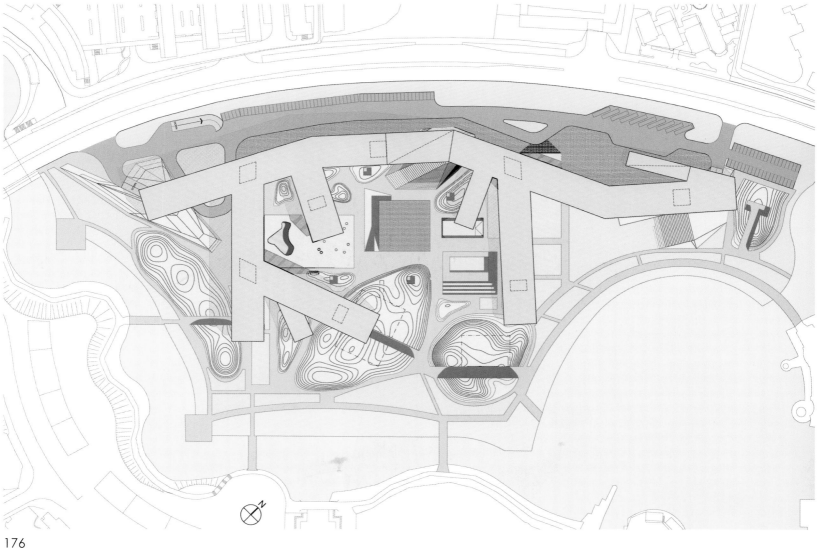

SIZE COMPARISON
规模对比

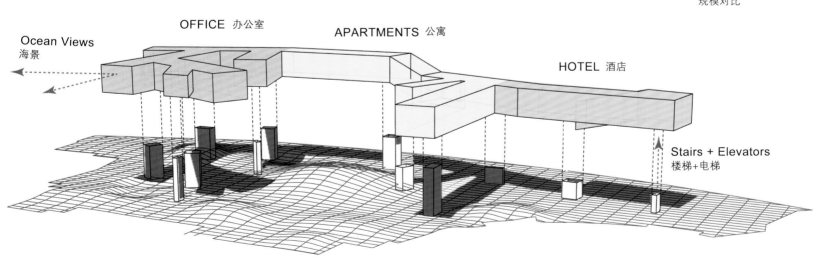

Ocean Views
海景

OFFICE 办公室

APARTMENTS 公寓

HOTEL 酒店

Stairs + Elevators
楼梯+电梯

LANDSCAPE 景观

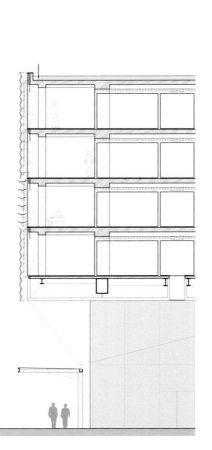
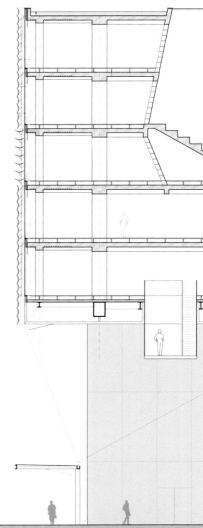

New Acropolis Museum
新卫城博物馆

Jury Comments:

The building rotates in plan to fit the site – it is very contextual and powerfully respectful of the urban fabric of Athens while doing a dance around the ruins.

Not a light building – it is very contextual and powerfully respectful of the urban fabric of Athens while doing a dance around the ruins.

The sculpture from the old museum is much more dramatic than in the old setting with the screen walls and slab edges remaining contextual to the neighborhood and city.

评委评语：

建筑通过旋转来适应场地——它充分考虑了多方环境，在雅典的城市网格内围绕着废墟而舞动。

这不是一座轻飘飘的建筑。

围墙和天花板边缘与城市环境的紧密联系让旧博物馆中的雕塑在新博物馆中更具戏剧效果。

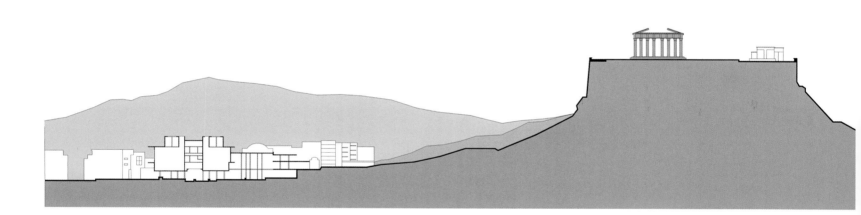

Notes of Interest

Located in Athens's historic Makryianni district, the New Acropolis Museum stands less than 1,000 feet southeast of the Parthenon. The site presented unique challenges, such as the need to accommodate a large existing structures and Athens' largest ongoing archaeological excavation. Additionally and importantly, the site is located in a hot climate in a major seismic zone, requiring state-of-the-art technology to protect visitors, staff, and the irreplaceable artifacts of the Museum collection.

The building was designed in three layers, two of which follow the city grid and existing and ancient pathways. The top-floor Parthenon Gallery, designed to display the Parthenon sculptures, is rotated 23 degrees and dimensioned to approximate the size, orientation, and viewing conditions of the historic Parthenon.

The Parthenon Gallery's glass outer walls allow visitors uninterrupted, 360-degree views of the ancient temple and the surrounding city. Its transparent enclosure provides ideal light for sculpture in direct view to and from the Acropolis, using the most contemporary glass and climate control technology, engineered with a view to sustainability, to protect the gallery against excessive heat and light. One of the goals of the topmost gallery is to eventually reunite the elements of the Parthenon Frieze, currently dispersed among several world museums.

Associate Architect: Michael Photiadis & Associate Architects
Owner: Greek Ministry of Culture
合作建筑师： 梅克尔·弗蒂亚迪斯建筑事务所
所有人： 希腊文化部

Architect / 建筑师
Bernard Tschumi Architects
伯纳德·曲米建筑事务所

Location / 项目地点
Athens, Greece
希腊，雅典

Photo Credit / 图片版权
© Christian Richters, Exterior; © Peter Mauss/Esto, Interior
克里斯汀·雷切特斯（外部）；彼得·茅斯/埃斯托（室内）

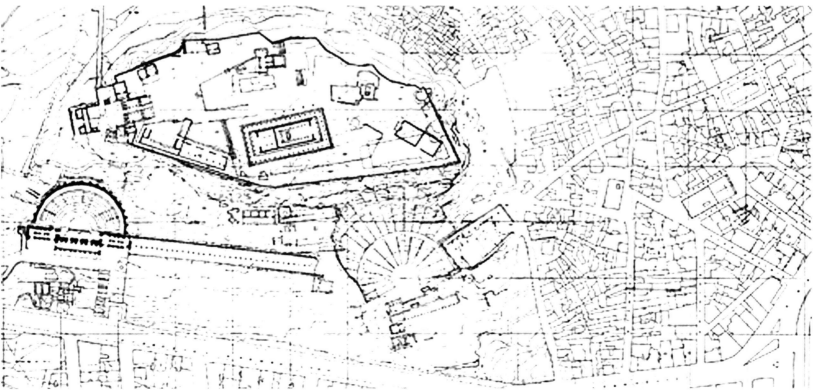

项目特色

新卫城博物馆位于雅典的马克里雅尼区,距离巴特农神殿仅300米。项目面临着独特的挑战。例如,它必须容纳一座大型历史建筑和雅典最大的进行中的考古遗址。更为重要的是,项目处在炎热的气候中,地处主要地震带,需要运用最先进的技术来保护游客、员工以及博物馆无与伦比的藏品。

建筑被分为三个层面,其中两层遵循了城市网格以及原始的古城路径。顶楼的巴特农画廊专门展示巴特农神殿的雕像,画廊旋转了23度,与古巴特农神殿的尺寸、方位以及视野条件都极为相似。

巴特农画廊的玻璃外墙让游客可以享有神殿和雅典城的360度无障碍美景。它的透明外壳为卫城的雕塑提供了完美的光线。项目利用最现代的玻璃和气候控制技术以及可持续工程技术来保护画廊不受多余的光热损坏。顶楼画廊的目标之一是逐渐重聚巴特农神殿的雕塑(目前它们散落在世界各地的博物馆中)。

North Carolina Museum of Art
北卡罗来纳艺术博物馆

Jury Comments:

Worthy of recognition for the precision and technology that went into the design of the ceiling and light well – the way daylight is brought into this building is ingenious.

From a distance, the building appears as a normal industrial building fitting into its context – upon approach it is an amazingly precise and elegant box.

Very unique for a museum in that it contributes to the overall master plan for this part of the city.

评委评语：

天花板和光井设计的精确度和技术值得称赞——阳光被创造性地引入建筑内部。

从远处看，建筑宛如嵌入环境的普通工业建筑；走近一看，它显得无比精密而优雅。

这是一个十分独特的博物馆，它的整体规划嵌入了城市的网格之中。

Notes of Interest

Inside the North Carolina Museum of Art, the light of day and the lush surrounding hills have a presence unusual in institutional galleries for art. Overhead, hundreds of elliptical occuli bathe the museum's interior in even, full-spectrum daylight, modulated in intensity by layered materials that filter out damaging rays.

A departure from tradition, the museum in some respects is a single 65,000-square-foot room. Within this spatial continuum, a succession of wall planes delineate separate galleries. The building's skin is a rain screen of pale, matte anodized-aluminum panels that softly pick up surrounding colors and movement, fostering a discourse with the landscape.

The expansion galleries at the North Carolina Museum of Art will provide a distinct visitor experience in a state-of-the-art "energy smart" building. Naturally illuminating the interior environment provides color rendering and light levels ideal for viewing art, while efficient temperature and air quality controls, lighting and envelope systems provide the ideal interior environment for preserving the art.

Engineer: Skidmore, Owings & Merrill LLP AltieriSeborWieber, Kimley-Horn Associates
General Contractor: Balfour Beatty, Barnhill
Lighting: Fisher Marantz Stone
Owner: Department of Cultural Resources, State of North Carolina

合作建筑师：皮尔斯·布林克利·西斯+李
工程师：SOM建筑事务所、ASW公司、金利–霍恩事务所
总承包商：巴尔弗·贝蒂、巴恩希尔
照明设计：费舍尔·马兰士·斯通
所有人：北卡罗来纳州文化资源部

Architect / 建筑师
Thomas Phifer and Partners
托马斯·费佛建筑事务所

Location / 项目地点
Raleigh, North Carolina
北卡罗来纳州，罗利

Photo Credit / 图片版权
© Scott Frances
斯科特·弗朗西斯

项目特色

在北卡罗来纳艺术博物馆内,人们可以享受阳光和周边青翠的小山景色,这在传统的美术馆中十分罕见。头顶上,成百上千的椭圆形天窗让博物馆内部充满了阳光,层叠的材料滤去了有害光线。

这座颠覆传统的博物馆总面积约6,040平方米。在连续的空间中,一系列墙板分划出一间间独立的展厅。建筑的表皮遮雨板由灰白的亚光镀锌板制成,与周边的色彩柔和地结合起来,和景观形成了对话。

北卡罗来纳艺术博物馆的扩建展厅将以最先进的"智能能源"策略为来访者提供超凡的体验。室内环境的自然采光提供了观看艺术品的色彩渲染和光线层次;高效的温度和空气质量控制、照明以及表皮系统为保护艺术品提供了理想的环境。

One Jackson Square
杰克逊一号广场

Jury Comments:
This project is textural and spatial to a high degree.
Great example of using modern digital fabrication techniques.
The execution of the window wall is a strong resolution of the detail nicely resolved.

评委评语:
项目在结构和空间上都达到了卓越的高度。
现代数码装配技术的运用典范。
窗口墙面的设计别出心裁,细部设计精美。

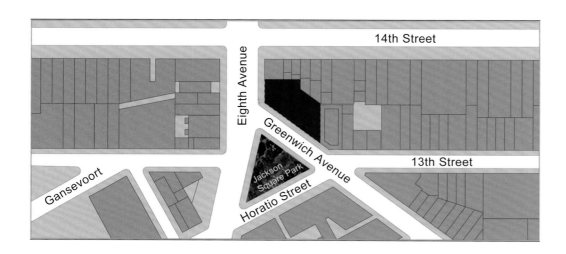

Notes of Interest

One Jackson Square, located in Manhattan's Greenwich Village, is a 35-unit luxury residential development that responds to its celebrated locale. This district is home to the highest concentration of early architecture in New York City, and any new structures introduced to this intricate fabric must respect its existing architecture and history.

Formerly a surface parking lot, the six-sided, split-zone site above two subway tunnels posed significant challenges, which the design negotiates through its massing, material expression, and robust foundation. It also provides a long-missing north edge to Jackson Square Park.

The building volume steps down from 11 stories to seven stories, from north to south, accommodating the zoning laws and mediating the varied scales of the neighborhood. Undulating bands of glass identify individual floors, creating a ribbon-like series of convexities and concavities along the street wall. The predominantly masonry structures of the immediate surroundings, along with the park, are "played back" in the glazed, fluid façade.

Associate Architect: Schuman Lichtenstein Claman Efron
Engineer: RA Consultants, WSP Flack & Kurtz, Gilsanz Murray Steficek
General Contractor: Hunter Roberts Construction Group
Historic Preservation: Higgins & Quasebarth
Owner: Hines

合作建筑师:SLCE公司
工程师:RA咨询公司、WSP弗莱克&库尔斯、杰尔桑斯·马雷·斯蒂菲克
总承包商:亨特·罗伯斯建筑集团
历史保护:希金斯&夸塞巴特
所有人:汉斯公司

Architect
Kohn Pedersen Fox Associates, PC
KPF 建筑事务所

Location
New York City, New York
纽约州，纽约

Photo Credit
© Raimund Koch
雷蒙德·科赫

Lower Level Section:
1. Roof terrace
2. Terrace
3. Duplex level 6
4. Apartment level 5
5. Apartment level 4
6. Apartment level 3
7. Apartment level 2
8. Retail
9. Basement

底座剖面图：
1. 屋顶平台
2. 平台
3. 六楼跃层公寓
4. 五楼公寓
5. 四楼公寓
6. 三楼公寓
7. 二楼公寓
8. 零售空间
9. 地下室

Jackson Square Park 杰克逊广场公园
Greenwich Avenue 格林威治大道
Eighth Avenue Subway 8号大道地铁

项目特色

杰克逊一号广场位于曼哈顿的格林威治村，是一座拥有35个单位的奢华住宅开发项目，专为当地的名人设计。这一区域遍布纽约城的早期建筑，任何新建筑结构都必须尊重当地现有的建筑和历史。

场地的前身是一个停车场，这个六边形分割区域坐落在两条地下管道之上，为项目带来了巨大的挑战。设计通过建筑体量、材料和坚实的地基解决了这一问题。设计还给杰克逊广场公园提供了北侧边缘。

建筑由北向南从11层下降到7层，以符合区域分区法规，并与周边区域不同规模的建筑融为一体。起伏的玻璃带界定了独立的楼层，在沿街墙面上形成了缎带一般的系列凸起和凹面。周边显赫的石造建筑结构与公园一起被倒映在玻璃外立面上。

San Francisco Museum of Modern Art Rooftop Garden
旧金山现代艺术博物馆屋顶花园

Jury Comments:
The notion of fitting this unique series of spaces together at a rooftop level and creating an interesting and exciting venue for the museum to sponsor is the hallmark of this project.
It is its own space and environment but aware of the city surrounding it.
This is a model of how we can enrich the urban fabric via a pavilion rooftop and the safety of a cloistered area.

评委评语:
项目的特色是在屋顶上添加独特的展览空间,为博物馆打造有趣而令人兴奋的空间。
独特的空间环境与周边的城市紧密结合。
项目是通过屋顶展厅和隐蔽空间来丰富城市结构的典范。

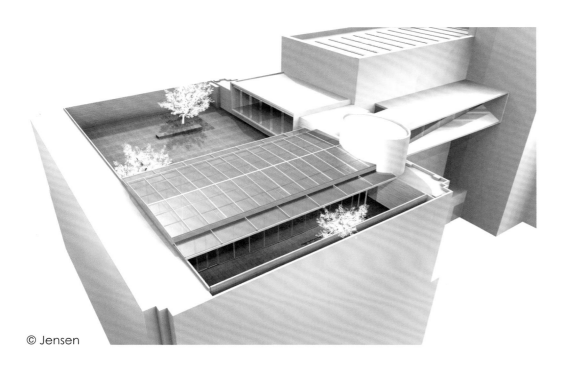
© Jensen

Notes of Interest
The San Francisco Museum of Modern Art Rooftop Garden was conceived as a gallery without a ceiling, defined by the intersection of sculpture, space and light, serving as a quiet, contemplative space for viewing art and hosting the museum's special events.

The garden spaces are accessed via a glass-enclosed bridge that affords sweeping views of downtown and the city's skyline. The new bridge provides circulation down its sloping floor towards the awaiting Pavilion, simultaneously adding an additional 1,500 square feet for art display.

In order to further integrate the Rooftop Garden within the sequence of existing galleries a 3,000-square-foot extension of the fifth floor gallery, suitably named the Overlook, was designed between the rear of the museum and the new garden. The entire back wall of the Overlook is glazed with a large panoramic window, allowing visual connection between gallery and garden.

Consultant: Charles M. Salter Associates, Shenyang Yuanda Aluminum Industry Engineering Co., Ltd
Engineer: Guttmann & Blaevoet Consulting Engineers, Forell / Elsesser Engineers, Inc.
General Contractor: Pdfdfdf
Lighting: Horton Lees Brogden Lighting Design
Owner: San Francisco Museum of Modern Art
顾问: 查尔斯·M·沙尔特事务所、沈阳远大铝业工程公司
工程师: 加特曼&布里沃特工程咨询公司、弗莱尔/埃尔塞瑟尔工程公司
灯光设计: 霍顿·里斯·布罗戈登照明设计
所有人: 旧金山现代艺术博物馆

Architect / 建筑师
Jensen Architects/Jensen & Macy Architects
詹森建筑事务所 / 詹森 & 梅西建筑事务所

Location / 项目地点
San Francisco, California
加利福尼亚州，旧金山

Photo Credit / 图片版权
© Bernard Andre Photography, © Richard Barnes Photography, © Henrik Kam Photography
伯纳德·安德烈摄影、理查德·巴恩斯摄影、亨里克·坎姆摄影

项目特色

旧金山艺术博物馆屋顶花园被设计成一个没有天花板的展览厅，通过雕塑、空间和光线的交错形成了一个宁静祥和的艺术空间，让人们在此观赏艺术品和举办博物馆的特殊活动。

人们通过一个玻璃天桥进入花园空间。在天桥里，人们可以将城市的景色一览无余。新桥通过下降的楼层直达等候大厅，为艺术馆增添了近140平方米的展览空间。

为了进一步将屋顶花园与原有的展览厅结合起来，设计师在五楼博物馆后部和新花园之间新增了278平方米的展览空间，名为"眺望"。"眺望"的整面后墙都是全景窗，在展厅和花园之间建立了视觉联系。

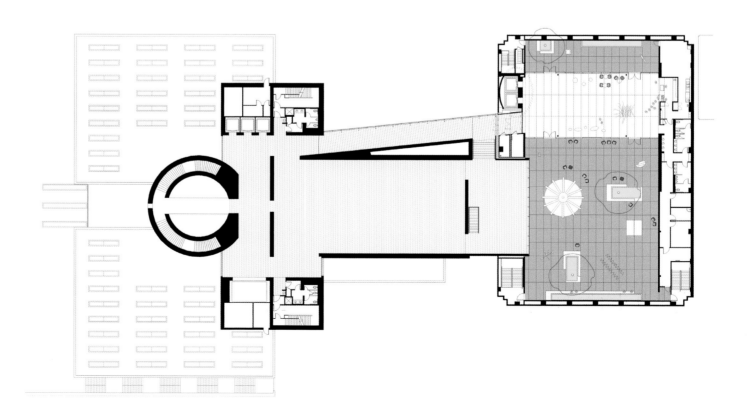

197

The Barnard College Diana Center
巴纳德学院戴安娜中心

Jury Comments:
This project is elegant and inviting. The integration of the landscape and architecture is the hallmark of this project and the heart of a good campus building. It still gives a sense of solidity to the wall between the enclosed campus and the outside world, but allows some transparency from the building out to Broadway and back in from the street.

评委评语：
项目优雅而吸引人。景观和建筑的结合是项目的特色，也是其成为优秀校园建筑的主要原因。项目为封闭校园和外界之间的墙壁注入了封闭感，同时又为建筑朝向百老汇和街面的一侧带来了通透感。

Section:
1. Academic departments
2. Conf/mto/semiar
3. Anth lab
4. Ta/conf
5. Arch studio
6. Senior proj/painting
7. Computer lab
8. Arch adj
9. Breakout
10. Students conf
11. Wac
12. Workroom
13. Office
14. Servery
15. Servery support
16. Storage
17. Control room
18. Events
19. Prefunction
20. Toilet
21. Mech

剖面图：
1. 学院部门
2. 会议室/研讨室
3. 实验室
4. 助教工作室
5. 建筑工作室
6. 高级项目室/绘图室
7. 计算机实验室
8. 建筑调整室
9. 休息室
10. 学生会议室
11. 网络控制室
12. 工作室
13. 办公室
14. 备餐间
15. 备餐间辅助室
16. 储藏室
17. 控制室
18. 活动室
19. 准备室
20. 洗手间
21. 机械室

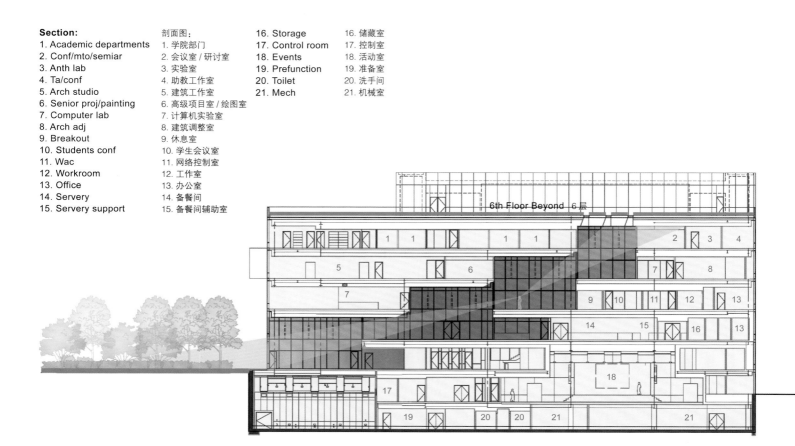

Notes of Interest

The Barnard College Diana Center's site is set within an intimate campus compressed within the dense, urban environment of Manhattan. Located between the Lehman Lawn and Broadway, the Diana Center unites landscape and architecture, as well as interior and exterior spaces, presenting a window onto the College and the city. The 98,000-square-foot multi-use building establishes an innovative nexus for artistic, social, and intellectual life on the campus. The facility brings together spaces for art, architecture, theater, and art history, as well as faculty offices, a dining room, and a Café.

From the historic entrance gate at Broadway, the wedge-shaped design frames a clear sightline linking the central campus at Lehman Lawn to the lower level historic core of the campus. The Diana Center extends Lehman Lawn horizontally and vertically; descending planted terraces cascade north to Milbank Hall, previously isolated by a 14 foot-high retaining wall plaza, and ascending double-height atria bring natural light and views into the seven story structure.

Consultant: Fisher Dachs, Jaffe Holden Acoustics
Engineer: Langan, Jaros, Baum & Bolles Consulting Engineers, Severud Engineers
Lighting: Brandston Partners, Inc.
Owner: Barnard College

顾问： 费舍尔·达奇斯、杰夫·霍顿音效设计
工程师： 兰恩、JBB工程咨询公司、瑟弗拉德工程公司
照明设计： 布兰德斯通公司
所有人： 巴纳德学院

Architect / 建筑师	Location / 项目地点	Photo Credit / 图片版权
Weiss/Manfredi Architecture/Landscape/Urbanism 韦斯 / 曼菲蒂建筑 / 景观 / 城市规划事务所	New York City, New York 纽约州，纽约	© Albert Vecerka/Esto, © Paul Warchol Photography 艾伯特·威斯尔卡 / 埃斯托、保罗·瓦尔孔摄影

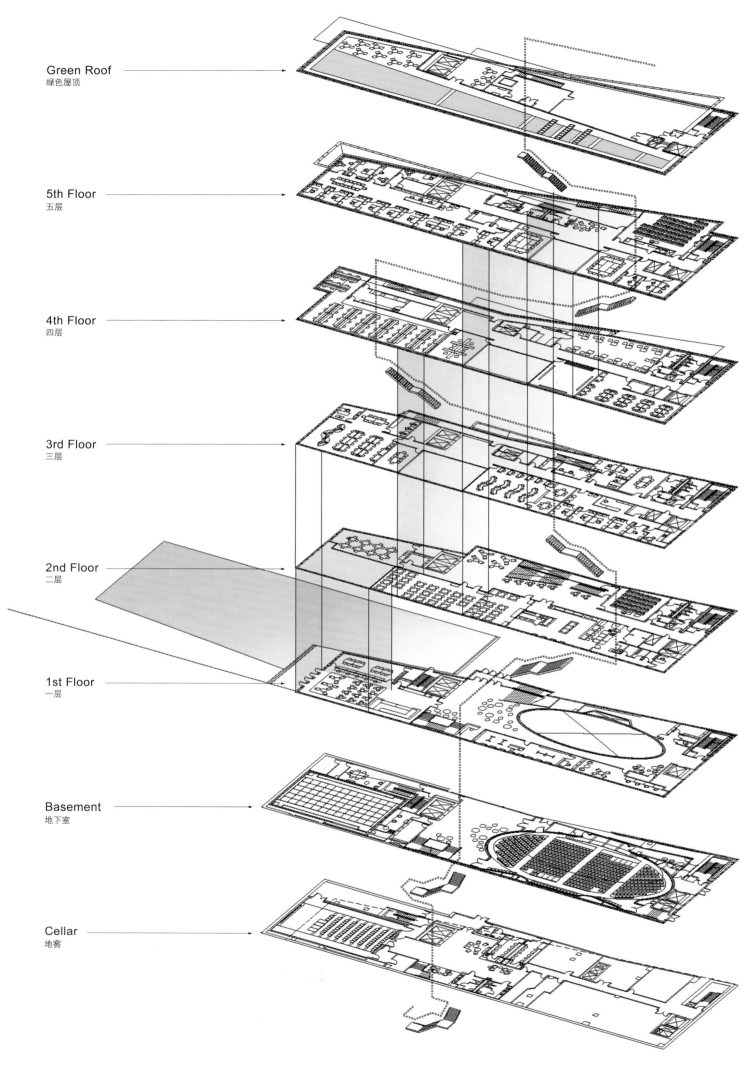

© Albert Vecerka/Esto

© Albert Vecerka/Esto

201

项目特色

巴纳德学院置身于曼哈顿密集的城市环境之中，戴安娜中心就坐落在这个私密的校园之中。戴安娜中心位于雷曼草坪和百老汇之间，将景观与建筑、室内与室外空间统一了起来，为学院和城市打开了一扇窗。这座9,104平方米的多功能建筑在校园的艺术、社交和学术生活之间建立的创意联系。中心集艺术、建筑、剧院和艺术历史空间于一身，同时还设置了办公室、餐厅和咖啡厅。

从百老汇一侧的历史入口进入，这个楔形设计在视觉上将雷曼草坪的中央园区与下方的校园历史中心连接了起来。戴安娜中心在水平和垂直方向上延伸了雷曼草坪；下降的绿色平台向北直达米尔班克厅，而上升的前庭则为建筑内部带来了自然光和良好的视野。

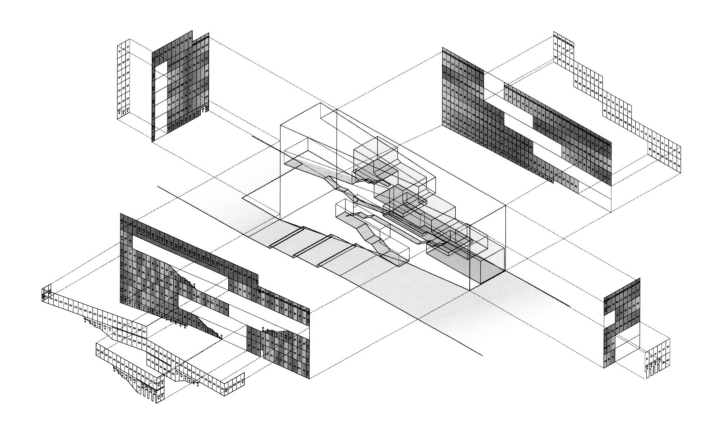

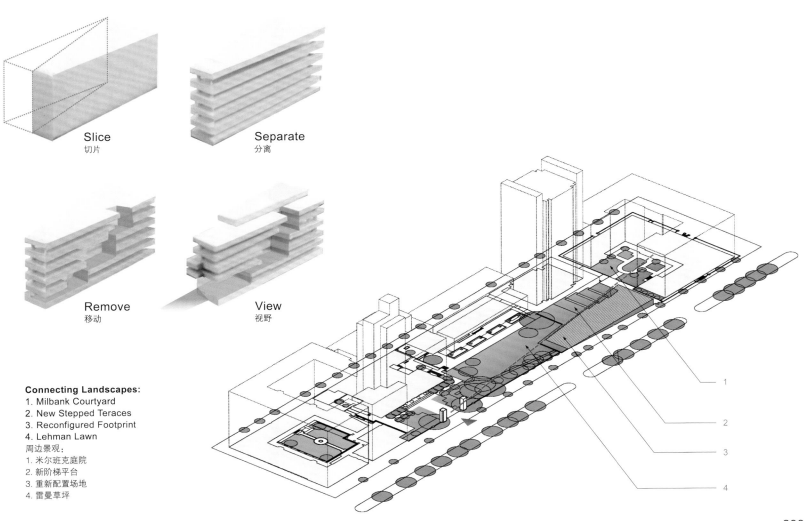

Slice
切片

Separate
分离

Remove
移动

View
视野

Connecting Landscapes:
1. Milbank Courtyard
2. New Stepped Teraces
3. Reconfigured Footprint
4. Lehman Lawn

周边景观：
1. 米尔班克庭院
2. 新阶梯平台
3. 重新配置场地
4. 雷曼草坪

University of Michigan Museum of Art
密歇根大学艺术博物馆

Jury Comments:
The prominence of the site and juxtaposition of the older bold with new bold is a strong symbol for the university protecting their past and looking towards the future.
A new addition to an existing historic building reads as a new building and true to itself.

评委评语:
场地的重要性与新旧建筑的对比体现了大学保护历史、展望未来的决心。
在历史建筑上新增的建筑结构宛如一座全新的建筑,真实而自信。

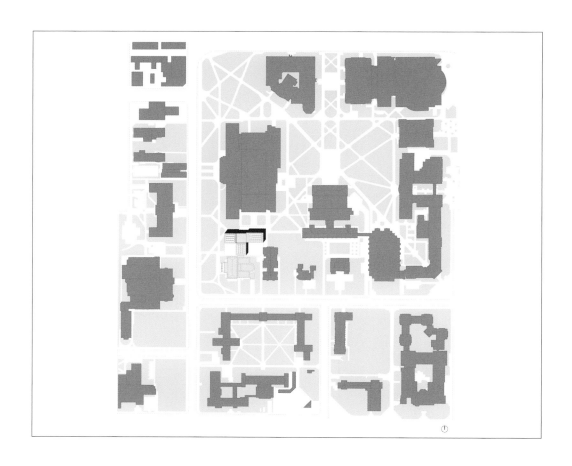

Notes of Interest

In creating the University of Michigan Museum of Art (UMMA), the purpose was to completely renovate and modernize the existing Alumni Hall and build an additional 53,452 square feet of space in a dramatic new wing. Located at the gateway to the University's main campus and at the physical intersection of the University and local communities, the Museum rests in a location offering the potential for direct engagement between the Museum, the student body and the general public.

As a teaching museum with broad, near universal collections, the institution serves as a forum for the various academic disciplines of the University as well as a cultural portal for the community of Ann Arbor. While the existing building provides an atmosphere of seclusion, the new architecture achieves an immediacy with the surrounding campus – inviting and even provoking engagement with the building and its programs.

Associate Architect: Integrated Design Solutions
Consultant: RA Heintges, SGH
Engineer: Atwell-Hicks, Arup, KPFF
General Contractor: Pdfdfdf
Landscape Architect: HAWA
Owner: University of Michigan
合作建筑师: 综合设计方案公司
顾问: RA·亨特吉斯、SGH
工程师: 亚特维尔-希克斯、奥雅纳工程顾问公司、KPFF
所有人: 密歇根大学

Architect / 建筑师
Allied Works Architecture
联合工作建筑事务所

Location / 项目地点
Ann Arbor, Michigan
密歇根州，安阿伯

Photo Credit / 图片版权
© Richard Barnes, Interior, © Jeremy Battermann, Exterior
理查德·巴恩斯（室内）、杰里米·巴特尔曼（外部）

项目特色

密歇根大学艺术博物馆的设计目的是完全修复原有的校友大厅并且在旁边新增一个4,965平方米的空间。博物馆坐落在大学校园的通道之上,是大学和当地社区的交叉点,为博物馆、学生群体和公共民众提供了相互沟通的机会。

作为一个拥有大量藏品的教学博物馆,它为大学各种各样的学术活动提供了论坛,也为安阿伯的社区提供了文化门户。原有的建筑拥有一种隐蔽的氛围,新建筑则与校园紧密联系,散发出吸引人的气息。

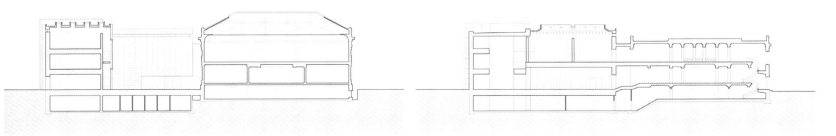

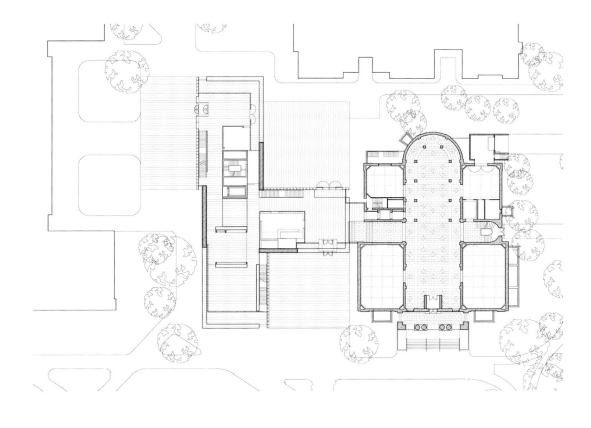
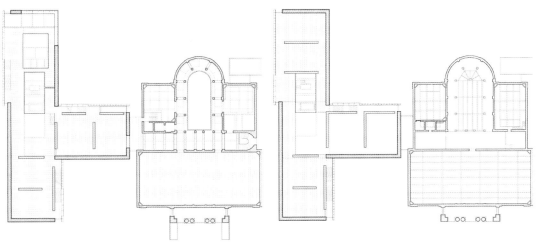

U.S. Land Port of Entry
美国陆上入境口岸

Jury Comments:
Elegantly premeditated, the building comes out of its function – the conflict between openness and security is pulled off well, while it also feels civic and like it belongs.
It is in a sense a continuation of the regional vocabulary – siding, wood, appropriate for its surrounds – integration of the landscape and public art – with the architecture nicely done.

评委评语：
通过优雅的设计，建筑实现了自身的功能价值——开放与安全之间的冲突被完美地解决，同时又呈现出民众归属感。
项目延续了区域元素——壁板、木材——对周边环境的尊重，结合了景观与公共艺术元素，打造了一座优秀的建筑。

Notes of Interest
The United States Land Port of Entry supports the mission-driven demands of Customs and Border Protection (CBP), the Department of Homeland Security's agency responsible for securing the nation's borders and promoting legal trade and travel.
Located in Warroad, Minnesota, this 43,000-square-foot facility is composed of three separate enclosed areas linked together with a continuous canopy. The port design manages a complex set of operational issues: the main building houses the officer work area and holding cells; the secondary building houses the vehicular inspection garages, laboratory space and firing range; and the commercial building is used for unloading and inspecting commercial vehicles. The port seamlessly integrates the latest technologies for securing the border into the facility and meets the demands of an energy-efficient and sustainable building.
In addition to meeting these programmatic and operational issues, the port also stands as a gateway to our nation, representing our open and democratic values of transparency, dignity, fairness and humaneness of our federal government.

Engineer: Jacobs, Sebesta Blomberg & Associates, Inc., Meyer Borgman and Johnson, Inc.
General Contractor: Pdfdfdf
Landscape Architect: Coen + Partners
Owner: GSA, Land Port of Entry Division
工程师： 雅各布斯、塞比斯塔·布罗姆伯格事务所、迈耶·伯格曼和约翰逊公司
总承包商： Pdfdfdf 公司
景观建筑师： 科恩事务所
所有人： GSA，美国陆上入境部

项目特色
美国陆上入境口岸支持美国海关与边防局任务需求——国土安全局负责保护国家边境和促进合法贸易和旅行的部门。
该项目位于明尼苏达州瓦路德，由三个独立封闭的区域组成。三者通过一个连续的顶棚连接起来。口岸设计解决了一系列复杂的运作问题：主楼内设有官员办公室区和拘留所；副楼是车辆检查车库、实验室和射击场；商业楼用作卸载和检查商业车辆。口岸设计结合了最先进的安全技术，符合节能和可持续建筑需求。
除了满足这些功能和运作需求，口岸还被作为美国的门户，呈现了美国联邦政府透明、高尚、公平和人性的开放民主价值观。

Architect / 建筑师
Julie Snow Architects, Inc.
朱莉·斯诺建筑公司

Location / 项目地点
Warroad, Minnesota
明尼苏达州,瓦路德

Photo Credit / 图片版权
© Paul Crosby
保罗·克罗斯比

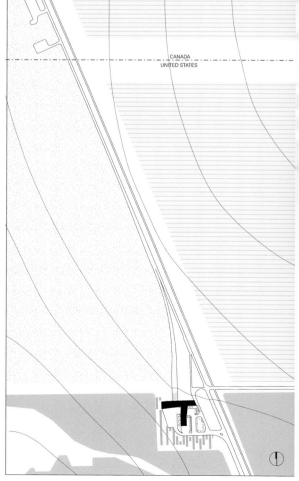

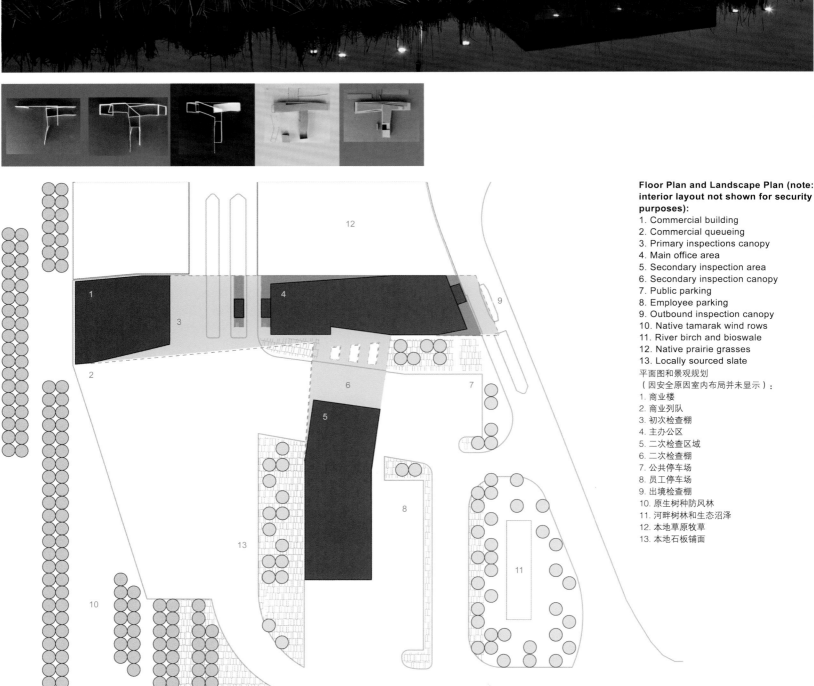

Floor Plan and Landscape Plan (note: interior layout not shown for security purposes):
1. Commercial building
2. Commercial queueing
3. Primary inspections canopy
4. Main office area
5. Secondary inspection area
6. Secondary inspection canopy
7. Public parking
8. Employee parking
9. Outbound inspection canopy
10. Native tamarak wind rows
11. River birch and bioswale
12. Native prairie grasses
13. Locally sourced slate

平面图和景观规划
（因安全原因室内布局并未显示）：
1. 商业楼
2. 商业列队
3. 初次检查棚
4. 主办公区
5. 二次检查区域
6. 二次检查棚
7. 公共停车场
8. 员工停车场
9. 出境检查棚
10. 原生树种防风林
11. 河畔树林和生态沼泽
12. 本地草原牧草
13. 本地石板铺面

Alchemist
炼金术师服装店

Jury Comments:
The design is respectful of the site's architecture but manages to shed the trappings of the conventional store by making its presence known in a subtly elegant and sophisticated manner.

评委评语:
设计在尊重场地上建筑的同时脱离了传统店铺的束缚，呈现出优雅而精致的视觉效果。

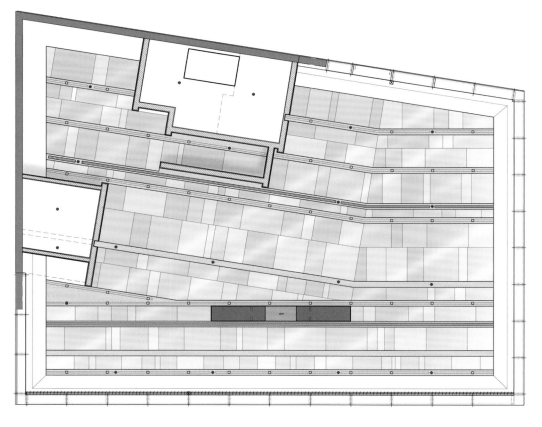

	Fire sprinklers 自动洒水系统
	Track lighting 活动式投射照明
	Ac diffusers 自动控制漫射灯

Site Elevation:
Scale: 1/16=1'-0"
场地立面图:
比例: 1/16=1'-0"

Engineer: Vidal & Associates, Optimus Engineering
General Contractor: Aaron Builders & Development
Lighting: Brand Lighting
Owner: Roma Cohen
工程师: 维达尔事务所、优驰工程公司
总承包商: 艾伦建筑开发公司
照明设计: 布兰德照明
所有人: 罗马·科恩

Notes of Interest

This sparkling glass box of retail is situated on the fifth-floor edge of a parking garage, yet somehow conquers impossible challenges: integrating the shop into the aesthetic of the parking structure and establishing connection and dialogue with the pedestrian environment below.

Notable is this project's ability to captivate both the store patrons and those meandering the streets of Miami Beach. It is composed of twenty-two foot high Starphire Glass which allows for expansive, crystal clear views into the space, allowing the play of light both inwardly and outwardly. The haunting transparency gives the project the appearance of being perched calmly like jewel box above the city. The choice of unobtrusive materials also provides a peaceful, fitting relationship between the shop and the larger parking structure.

In addition to this transparent glass, a complex system of mirrored walls and ceilings provide a dialogue of reflection between the store's goings on and the street below. They are interactive mirrors, operated on sensors which ripple in sync with the occupants' actions within the shop, generated by motion sensors and preset animations.

项目特色

这个闪闪发光的玻璃盒子造型服装店坐落在一个车库的五楼边缘，其设计面临着巨大的挑战：店铺需要融入停车场结构之中并且与下方的步行环境建立连接和对话。

值得注意的是项目既能够吸引老顾客，又能够吸引迈阿密海滩街头的游客。它由6.7米高的斯塔芬玻璃自创，为空间带来了宽阔而透亮的视野，让光线自由穿梭在室内外。通透感让项目宛如一个珠宝盒一样悬浮在城市顶端。低调材料的运用在店铺和更大的停车场结构建立了和谐的联系。

除了透明玻璃之外，一个复杂的镜面墙面和天花板系统倒映出店内和街道上的景象。这些交互的镜子通过传感器，随着店内人的行动而形成涟漪。

Architect / 建筑师
Rene Gonzalez Architect
雷内·冈萨雷斯建筑事务所

Location / 项目地点
Miami Beach
迈阿密海滩

Photo Credit / 图片版权
© Michael Stavaridis
迈克尔·斯塔瓦里迪斯

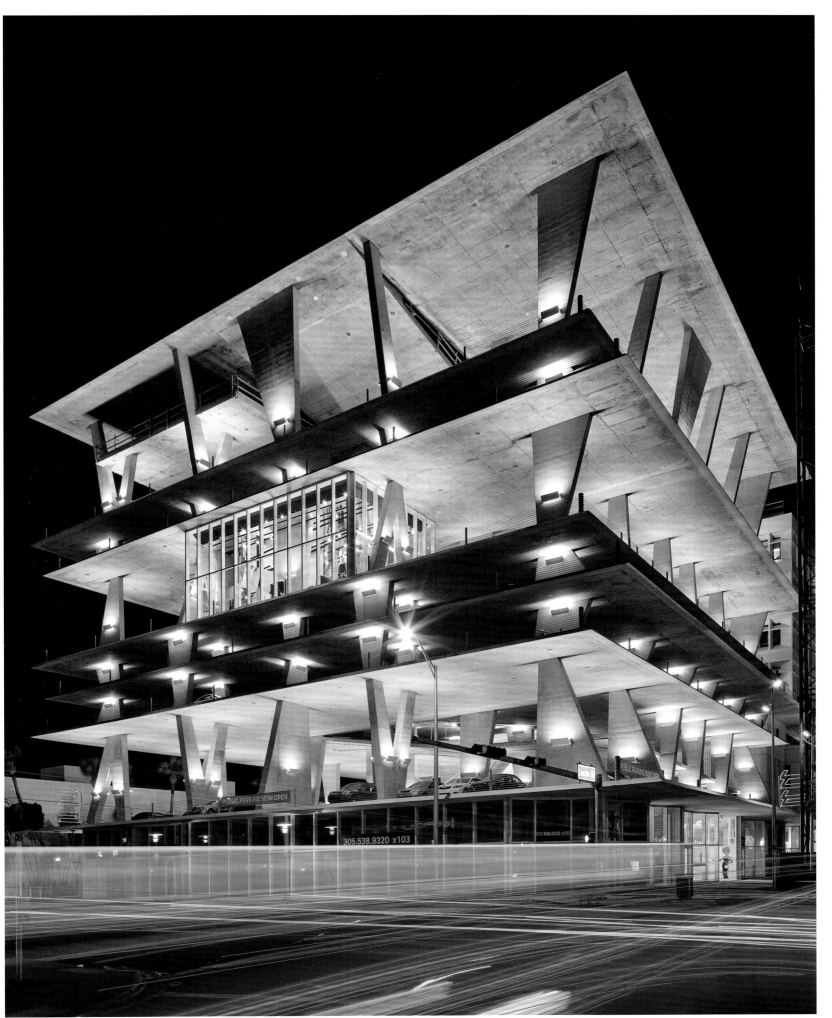

Site Plan+Location Image (Below): 总平面图+位置图像（下图）：
Scale: 3/32"=1'-0" 比例：3/32"=1'-0"
Location: Miami Beach 位置：迈阿密海滩
Climate Type: Humid Subtropical 气候类型：亚热带潮湿气候

Elevation (Below): 立面图（下图）：
1. Project level 1. 项目层
Elevation: 58'-4" 立面：58'-4"
2. Ground level 2. 地面层
Elevation: 0'-0" 立面：0'-0"

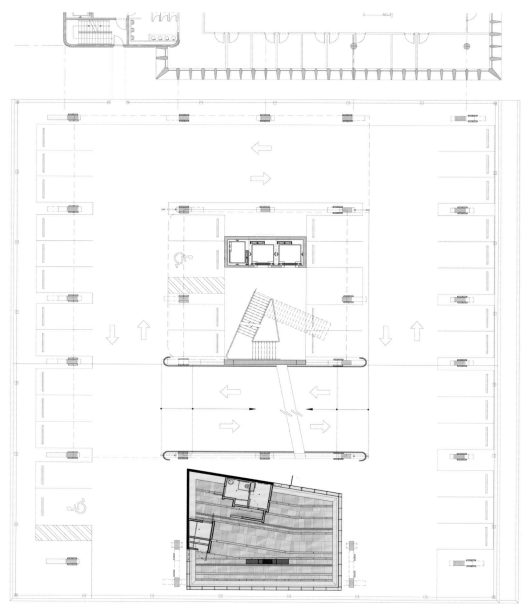
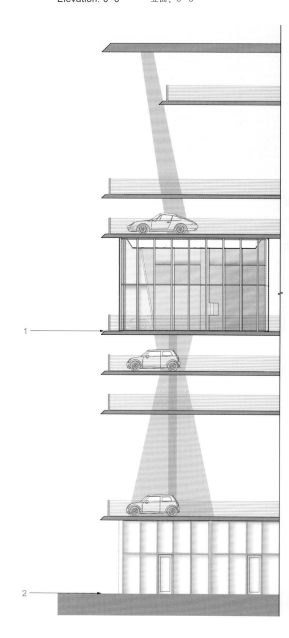

Armstrong Oil and Gas
阿姆斯特朗油气公司

Jury Comments:
Here, understated materials achieve elegance through superior detailing and craftsmanship.
This design stands out for its thoughtful space-making and through its handling of materials thoughtfully-chosen to respond to the character of the original building.
This project's expression of the best of what the original machine shop building had to offer is superbly celebrated with the architecturally honest palate of brick, steel, concrete and glass.

评委评语：
朴素的材料在这里通过卓越的细节设计和工艺散发出优雅的气息。
设计以其精心的空间布局而脱颖而出，它对材料的把握十分到位，与建筑的原始风格相互呼应。
项目展示了原始机械商店最好的一面，突出了砖石、钢铁、混凝土和玻璃的建筑价值。

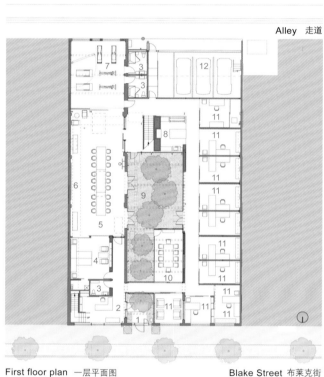

Site Plan (Left):
1. Entry Court
2. Reception
3. Restroom
4. Seismic Room
5. War Room
6. Catwalk
7. Fitness Lounge
8. Copy Room
9. Courtyard
10. Conference Room
11. Private Office
12. Garage
13. Lounge
14. Kitchen
15. Penthouse Office
16. Roof Terrace

总平面图（左图）：
1. 入口庭院
2. 前台
3. 洗手间
4. 地震安全室
5. 作战指导室
6. 天桥
7. 健身休息室
8. 复印室
9. 庭院
10. 会议室
11. 私人办公室
12. 车库
13. 休息室
14. 厨房
15. 阁楼办公室
16. 屋顶平台

First floor plan 一层平面图　　Blake Street 布莱克街
Second floor plan 二层平面图

Notes of Interest

This adaptive re-use of an early-1900's industrial machine shop launches a new identity for an established local business in lower downtown Denver. Charged with bringing new life to an underutilized building, the design team planned the enclosed program around existing elements in place and created generous, sophisticated spaces filled with daylight, natural ventilation and views to the Denver skyline.

In keeping with the historic manufacturing roots of the building, the structural steel is architecturally expressed throughout the building. Tipping a hat to the original materials, a firm contrast was maintained between their rustic, shell-blasted feel and the sharper, painted look of the newer elements.

To bring life to the space, the renovation brings new levels of circulation and transparency. The introduction of an interior courtyard sends daylight throughout the entire space, and translucent materials separating many of the workspaces capitalize on this natural light while balancing an element of privacy. The new office building consists of two main volumes and includes a breezeway, a conference room, a waiting area, an employee lounge, an open-air bridge, a roof terrace, outdoor meeting spaces, entertaining spaces and a beautifully redone penthouse office.

Architect of Record: Bothwell Davis George Architects, Inc.
Engineer: McGlamery Structural Group, M.E. Group
General Contractor: Sprung Consultant
Lighting: Fisher Marantz Stone
Owner: Armstrong Oil and Gas

记录建筑师：波斯维尔·戴维斯·乔治建筑公司
工程师：麦克格拉米利建筑集团、M.E.集团
总承包商：斯普朗咨询公司
照明设计：费舍尔·马兰士·斯通
所有人：阿姆斯特朗油气公司

Architect / 建筑师	Location / 项目地点	Photo Credit / 图片版权
Lake | Flato Architects 雷克|弗拉托建筑事务所	Denver, Colorado 科罗拉多州，丹佛	© Frank Ooms Photography 弗兰克·奥姆斯摄影

Building section through war room looking east:
1. Reception
2. Seismic Room
3. War Room
4. Fitness Lounge
5. Copy Room
6. Private Office
7. Garage
8. Lounge
9. Kitchen
10. Penthouse Office
11. Roof Terrace

从作战指导室看建筑东侧剖面：
1. 前台
2. 地震安全室
3. 作战指导室
4. 健身休息室
5. 复印室
6. 私人办公室
7. 车库
8. 休息室
9. 厨房
10. 阁楼办公室
11. 屋顶平台

Building section through war room looking east 从作战指导室看建筑东侧剖面

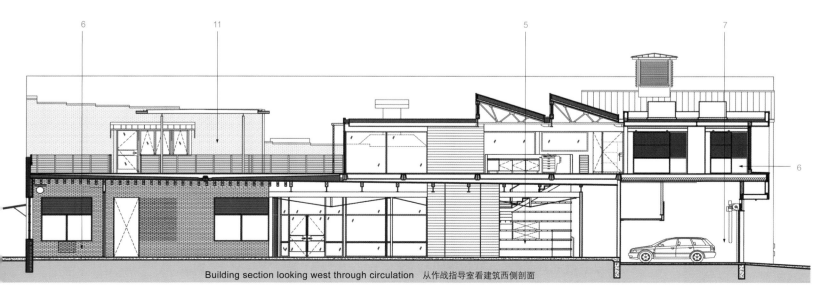

Building section looking west through circulation 从作战指导室看建筑西侧剖面

项目特色

项目将丹佛市中心一家建于20世纪初的机械商店改造成为当地知名公司的办公地点。设计团队为旧建筑带来了新生命,利用现有元素打造了宽敞而精致的空间环境,融入了自然采光、自然通风和良好的视野。

为了与建筑的机械制造业背景相一致,建筑内部大量使用了结构钢材。新旧材料形成了鲜明对比,纯朴的原始感与锋利的上色感相互矛盾。

为了给空间带来生机,翻新工作带来了全新的流通路径层次和通透感。内部庭院的使用为整个空间带来了自然光,隔开各个空间的半透明材料则完美地平衡了私密感。新办公楼由两个空间组成,其中包括带顶连廊、会议室、等候区、员工休息室、露天天桥、屋顶平台、户外集会空间、娱乐空间和阁楼办公室。

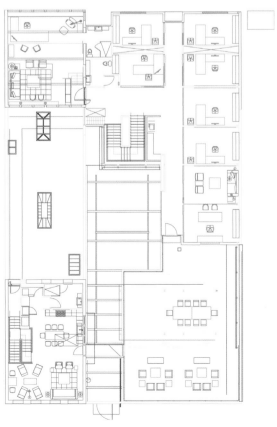

Conga Room
康加舞厅

Jury Comments:
The exploration of dance concepts, modular shapes, acoustics and bold colorful form exhibited in this project are combined in a way that one would never expect.
The innovation is full of analogy, sensuality, and technical directness that is just remarkable.

评委评语:
项目对舞蹈概念、模块造型、音效和大胆的色彩形式的探索营造出一种无以伦比的效果。
设计创新充满了比喻、感官和技术,令人难忘。

Notes of Interest

This newly constructed Los Angeles Latin live music venue is the premier location for Salsa and Rumba dancing in the area, and it was crucial that the space provide the advanced sound capabilities necessary to respond accordingly for its performers and patrons. But the challenge went even further: the dance hall had to be acoustically absorptive and isolated from the rest of the office building tenants while delivering this rewarding musical experience.

While facing these acoustical issues, the design solution also involves a dramatic visual experience, floating above a sea of dancers, a ceiling that acts as a very present character and that morphs in shape throughout the space while employing acoustical isolation and prescribed sound absorption. Additionally, in an effort to meet the client's aesthetic desires for a ceiling that reflected the vibrancy and dynamism of the Latin culture, a pattern made of diamonds was developed, inspired by the very dance step of Rumba itself!

One last notable flourish was the installation of a twenty-foot-tall lit "Tocado" – or headdress – draped from the ceiling of the primary entranceway. This captivating light display serves as a literal beacon of light, drawing attention from the plaza to the club.

Consultant: John Martin & Associate, John Dorius & Associate, A & F Consulting Engineers
General Contractor: Winters-Schram Associates
顾问: 约翰·马丁事务所、约翰·多利尔斯事务所、A & F工程咨询公司
总承包商: 温特斯–施拉姆事务所

Architect / 建筑师
Belzberg Architects
贝尔兹伯格建筑事务所

Location / 项目地点
Los Angeles, California
加利福尼亚州，洛杉矶

Photo Credit / 图片版权
© Benny Chan/Fotoworks
本尼·陈摄影

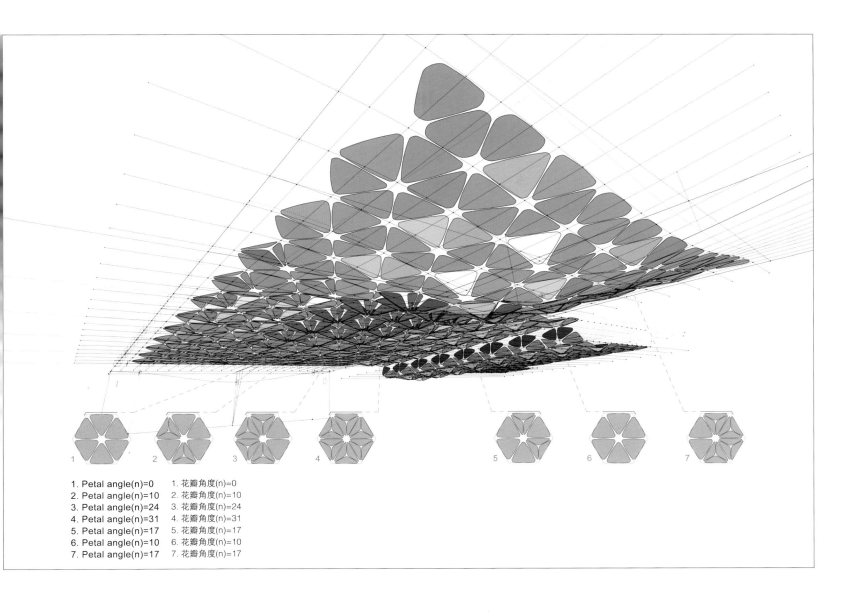

1. Petal angle(n)=0
2. Petal angle(n)=10
3. Petal angle(n)=24
4. Petal angle(n)=31
5. Petal angle(n)=17
6. Petal angle(n)=10
7. Petal angle(n)=17

1. 花瓣角度(n)=0
2. 花瓣角度(n)=10
3. 花瓣角度(n)=24
4. 花瓣角度(n)=31
5. 花瓣角度(n)=17
6. 花瓣角度(n)=10
7. 花瓣角度(n)=17

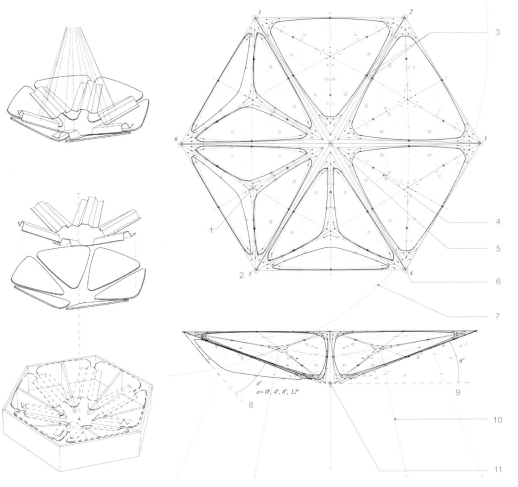

项目特色

这个新建成的洛杉矶拉丁音乐舞厅是当地主要的莎莎舞和伦巴舞舞厅。因此，舞厅必须提供先进的音响效果来呼应表演者和宾客的需求。但是设计还有另一个挑战：屋顶必须具有吸音功能、与办公楼的其他部分隔开，同时又要传递优秀的音乐体验。

在满足这些音效需求的同时，设计还打造了夸张的视觉体验。悬浮在舞者的海洋之上的天花板极具现代特色，随着空间的变换而不断变形，内部还嵌入了隔音和吸音设备。此外，为了满足委托人对天花板的美学要求，反映拉丁文化的活力，设计师从伦巴舞中获得了灵感，开放出了一种菱形图案。

最后，值得注意的是设计安装了一个6米高的顶灯，它从天花板一直垂到主入口。这个迷人的顶灯仿佛光线的灯塔，吸引了广场上的目光。

Elevation of Flower (Left):
1. Petal Radius(0) see below
2. Plan View Flower
3. Panel Offset Point
4. Petal Radius 1
5. Flower Radius(n) see below
6. Instantiation Points (1-7)
7. Petal Radius 1
8. o=0°, 4°, 8°, 12°
9. n=0°, 10°, 17°, 24°, 31°
10. Petal Surface Normals
11. Angle(n) controls the porosity of a flower based on acoustic

花朵立面（左图）：
1. 花瓣半径 (0)，见下图
2. 花朵平面图
3. 面板偏移点
4. 花瓣半径1
5. 花朵半径(n)，见下图
6. 偏移点(1-7)
7. 花瓣半径1
8. o=0°, 4°, 8°, 12°
9. n=0°, 10°, 17°, 24°, 31°
10. 标准花瓣表面
11. 角度(n)以声音为基础，控制花瓣的孔隙

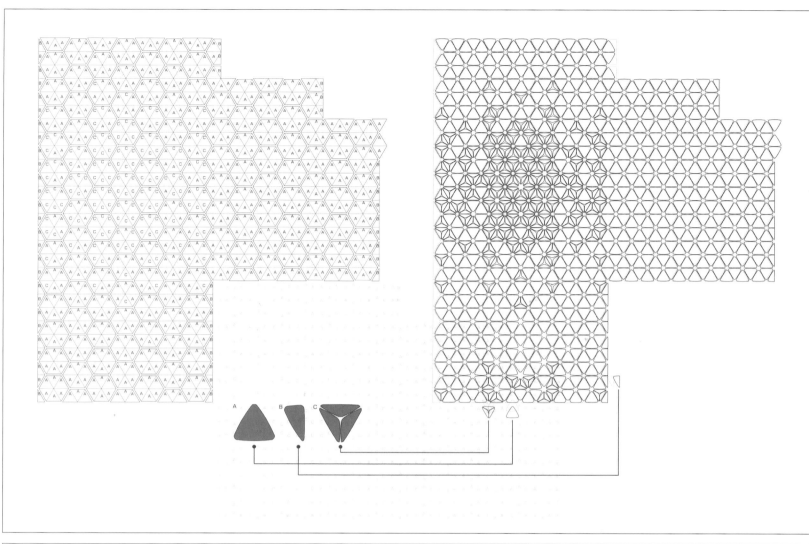
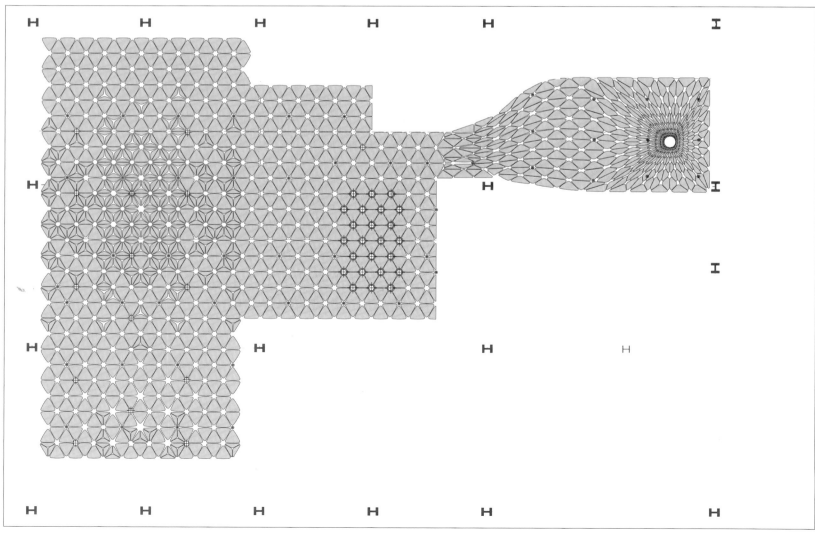

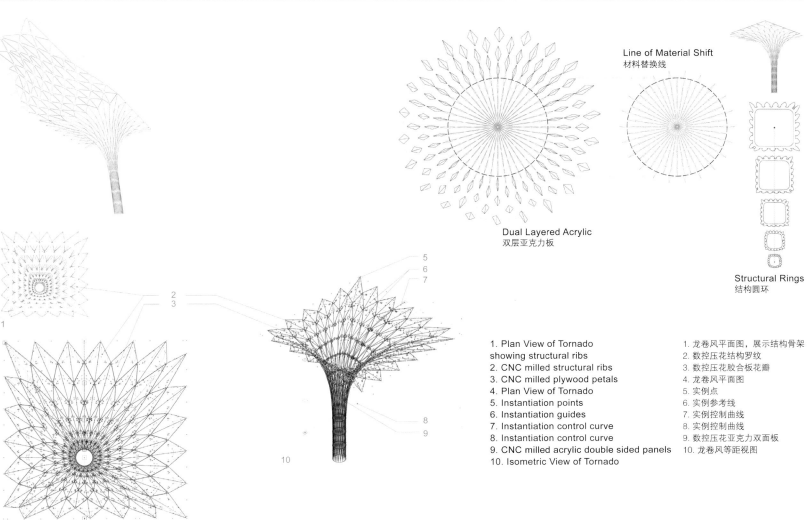

Dual Layered Acrylic
双层亚克力板

Line of Material Shift
材料替换线

Structural Rings
结构圆环

1. Plan View of Tornado showing structural ribs
2. CNC milled structural ribs
3. CNC milled plywood petals
4. Plan View of Tornado
5. Instantiation points
6. Instantiation guides
7. Instantiation control curve
8. Instantiation control curve
9. CNC milled acrylic double sided panels
10. Isometric View of Tornado

1. 龙卷风平面图，展示结构骨架
2. 数控压花结构罗纹
3. 数控压花胶合板花瓣
4. 龙卷风平面图
5. 实例点
6. 实例参考线
7. 实例控制曲线
8. 实例控制曲线
9. 数控压花亚克力双面板
10. 龙卷风等距视图

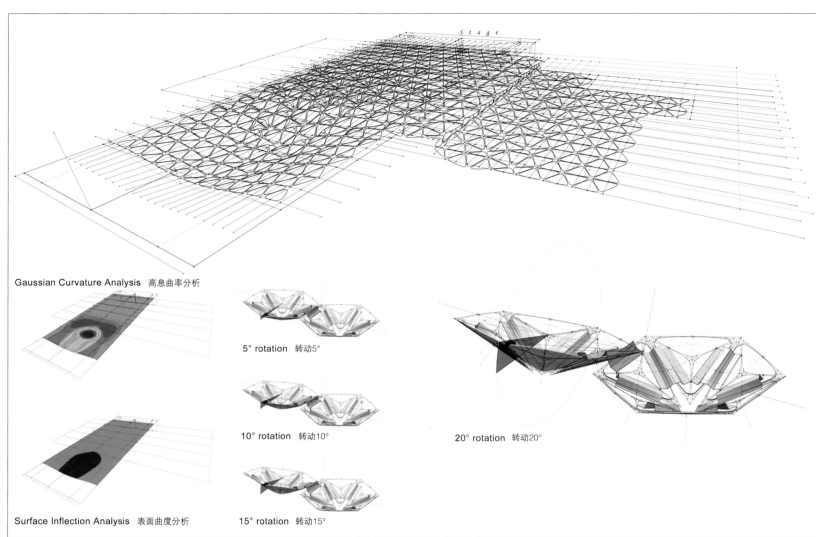

Gaussian Curvature Analysis 高息曲率分析

Surface Inflection Analysis 表面曲度分析

5° rotation 转动5°
10° rotation 转动10°
15° rotation 转动15°
20° rotation 转动20°

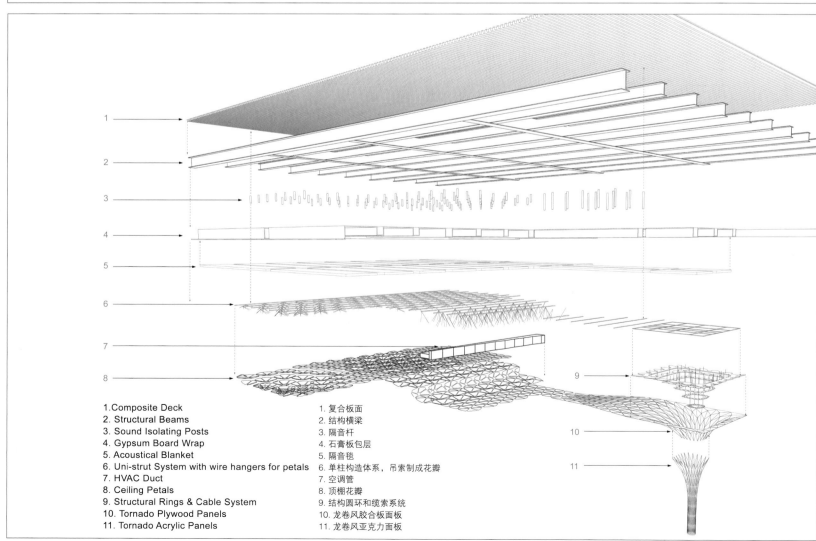

1. Composite Deck — 1. 复合板面
2. Structural Beams — 2. 结构横梁
3. Sound Isolating Posts — 3. 隔音杆
4. Gypsum Board Wrap — 4. 石膏板包层
5. Acoustical Blanket — 5. 隔音毯
6. Uni-strut System with wire hangers for petals — 6. 单柱构造体系，吊索制成花瓣
7. HVAC Duct — 7. 空调管
8. Ceiling Petals — 8. 顶棚花瓣
9. Structural Rings & Cable System — 9. 结构圆环和缆索系统
10. Tornado Plywood Panels — 10. 龙卷风胶合板面板
11. Tornado Acrylic Panels — 11. 龙卷风亚克力面板

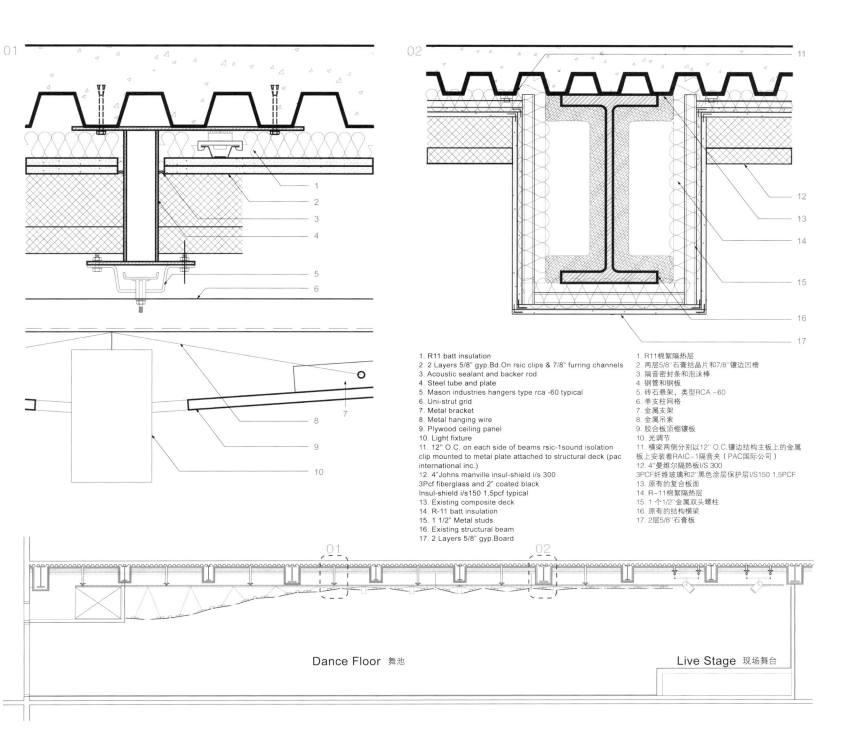

1. R11 batt insulation
2. 2 Layers 5/8" gyp.Bd.On rsic clips & 7/8" furring channels
3. Acoustic sealant and backer rod
4. Steel tube and plate
5. Mason industries hangers type rca -60 typical
6. Uni-strut grid
7. Metal bracket
8. Metal hanging wire
9. Plywood ceiling panel
10. Light fixture
11. 12" O.C, on each side of beams rsic-1sound isolation clip mounted to metal plate attached to structural deck (pac international inc.)
12. 4"Johns manville insul-shield i/s 300 3Pcf fiberglass and 2" coated black Insul-shield i/s150 1,5pcf typical
13. Existing composite deck
14. R-11 batt insulation
15. 1 1/2" Metal studs
16. Existing structural beam
17. 2 Layers 5/8" gyp.Board

1. R11棉絮隔热层
2. 两层5/8"石膏结晶片和7/8"镶边凹槽
3. 隔音密封条和泡沫棒
4. 钢管和钢板
5. 砖石悬架，类型RCA –60
6. 单支柱网格
7. 金属支架
8. 金属吊索
9. 胶合板顶棚镶板
10. 光调节
11. 横梁两侧分别以12" O.C.镶边结构主板上的金属板上安装着RAIC–1隔音夹（PAC国际公司）
12. 4"曼维尔隔热板I/S 300 3PCF纤维玻璃和2"黑色涂层保护层I/S150 1,5PCF
13. 原有的复合板面
14. R-11棉絮隔热层
15. 1个1/2"金属双头螺柱
16. 原有的结构横梁
17. 2层5/8"石膏板

Dance Floor 舞池   Live Stage 现场舞台

FIDM San Diego Campus
时装设计商业学院圣地亚哥校区

Jury Comments:
A playful, inspiring environment is created through skillful use of color and scale.
What might otherwise be a repetitious or simple environment is differentiated by unique spatial moments and interesting use of materials.
There is a cohesion to the space—despite the use of a wide variety of materials—due to the way the design interlocks the elements: the use of space defining space, no corridors, pockets of color keyed to a function, and lots of transparency moving light around the events, creates a learning environment that is what it espouses... design.

评委评语：
巧妙的色彩和比例打造了一个有趣而充满灵感的环境。
独特的空间和有趣的材料运用让原本重复或简单的环境大放异彩。
尽管设计采用了多种材料，由于各个元素之间的相互连接，项目空间保持了凝聚感。设计用空间定义空间、不设走廊、同一功能区采用同一色调并在活动区运用了许多透明移动光线，打造了一个与设计相匹配的学习环境。

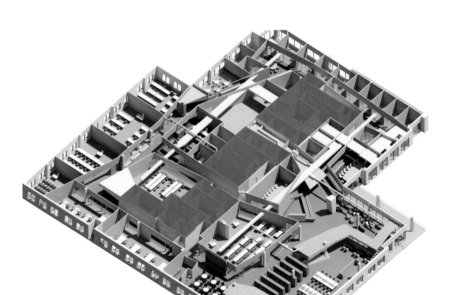

Notes of Interest
The Fashion Institute of Design and Merchandising (FIDM) is a private college with a thirty-five-year history of offering degrees directed at placing students in fashion, design and business. The college requested that its San Diego campus represent the school's progressive attitude towards education. The result is a space that is both non-traditional and tangibly centered around the value of design, appropriate enough for a school with just such a focus.

The project occupies the entire third floor of a high-rise office building. Comprising approximately 30,000 square feet, the space needed to accommodate all of the elements of the school's main campus within the smaller footprint of a regional campus. To achieve this, the school is designed as a sequence of zones: a public entry zone; an educational zone housing classrooms, the library, and technology resources; and an administration zone for the school's staff.

A looped circulation path encircles the floor plan, and generous public areas and hallway lounge settings create opportunities for spontaneous interaction. A strong color palette drawn from the area's native vegetation appears throughout the space. Additionally, a comprehensive graphic program that is integrated with the architecture connotes the function of spaces and leads users through the floor. While each area is self-defined through its color and form, integration between the spaces is very strong throughout.

Engineer: KPFF Consulting Engineers, Walsh Engineers
General Contractor: Steiner Construction
Owner: Fashion Institute of Design & Merchandising
工程师： KPFF工程顾问公司、沃尔什工程公司
总承包商： 斯坦纳建筑公司
所有人： 时尚设计商业学院

Architect / 建筑师
Clive Wilkinson Architects
克莱夫·威尔金森建筑事务所

Location / 项目地点
San Diego, California
加利福尼亚州,圣地亚哥

Photo Credit / 图片版权
© Benny Chan/Fotoworks
本尼·陈摄影

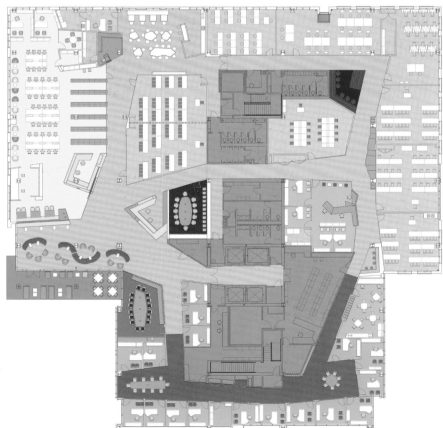

项目特色

时装设计商业学院是一家拥有35年历史的私立院校，专门培养时装、设计、商业方面的人才。学院要求圣地亚哥校区体现自身的教育进取精神。最终，设计师打造了一个非传统而又以设计价值为中心的空间，十分符合学院的中心价值。

项目占据了一座高层办公楼的整个三楼空间。空间总面积约2,787平方米，需要配置与学院主校区相同的全套设施。为了实现这一目标，学院被设计成一系列连续的空间：公共入口区、教学区（设有教室）、图书馆、技术资源区和教职员工行政区。

环形走道环绕着整个楼面，宽敞的公共区域和门厅设置打造了自然的互动机会。遍布整个空间的强烈色彩搭配来自于当地原生植物的外观。此外，项目的综合平面设计标志了各个空间的功能并引导着使用者。每个区域都有其独特的色彩和造型，空间之间的整合也贯穿始终。

Moving Picture Company
移动图形公司

Jury Comments:
The space is deftly crafted to be appropriate to the program and users it serves. The spatial sequence is cinematic – frame by frame, incorporating the various moods, lighting environments, theatrical clues, and subtle suggestion of movement through a process.
Thought and investigation went into this space that truly embodies themes of the user's business. Movement, drama, and light give this space a fun and dynamic feel.

评委评语:
空间的巧妙设计让人们感到无比舒适。空间序列如放映的影片一般——逐帧播放，融汇了各种各样的情绪、灯光环境、戏剧效果和运动暗示。
设计师的巧妙构思和调查研究让空间真实地展现了公司的专业主题。运动、戏剧和灯光让空间充满了趣味性和活跃感。

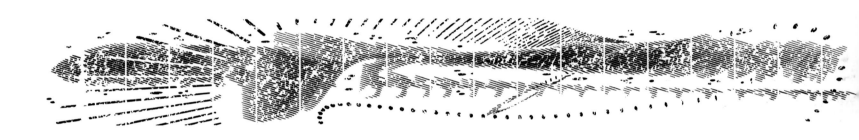

Notes of Interest

This 8,200-square-foot visual effects post-production facility is located within a generic office building in downtown Santa Monica, California. The Moving Picture Company is a United Kingdom-based visual effects post-production company, a forerunner in the visual effects and animation fields for the feature film, advertising, music and television industries. The facility serves as the United States Headquarters. Without the option of expanding in space further into the office building, the project needed to house grading rooms, edit bays, conference rooms, open and closed offices, client areas, production spaces, entertaining areas, tape vaults, mechanical rooms, machine rooms and support spaces.

An organic, sinuous spine weaves its way through the suite. An appendaged soffit grows from the serpentine walls and serves as an armature for cable trays, mechanical and electrical systems. Light portals pierce the organic forms and are equipped with programmable LED lighting. In addition to the provision of light, structure, and motion, the building itself serves as a reminder and homage to the professional field it aids.

Owner: The Moving Picture Company (MPC)
所有人: 移动图形公司

项目特色

这个总面积约760平方米的视觉效果后期制作公司坐落在圣塔莫尼卡一座普通的办公楼里。移动图形公司源于英国，是故事片、广告片、音乐片和电视产业等领域视觉和动画效果制作的先驱。这一项目是该公司的美国总部。项目无法扩展办公楼空间，需要将分级室、剪辑室、会议室、开放和封闭式办公室、客户区、制作空间、娱乐区、胶带储存室、机械室、机房和辅助空间汇聚在一起。

一个有机的蜿蜒走道将各个空间连接在一起。曲折的墙壁上凸出来一个附加的拱腹，被用作放置电缆架、机械和电力设备。进光口穿过有机造型的墙壁，并被辅以LED灯光照明。除了灯光、结构和运动感之外，建筑自身也充分体现了公司的专业个性。

Architect /建筑师
Patrick Tighe Architecture
帕特里克·泰伊建筑公司

Location /项目地点
Santa Monica, California
加利福尼亚州，圣塔莫尼卡

Photo Credit /图片版权
© Art Gray Photography
格雷艺术摄影

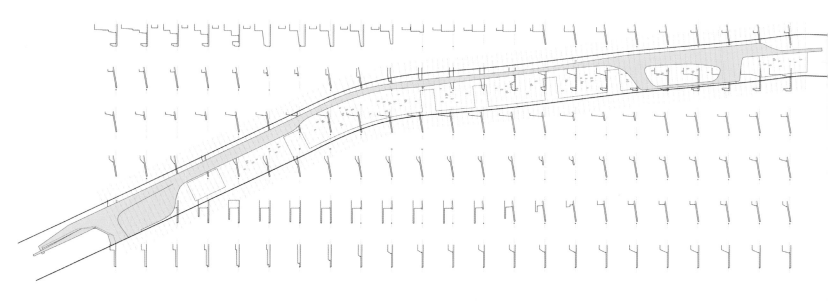

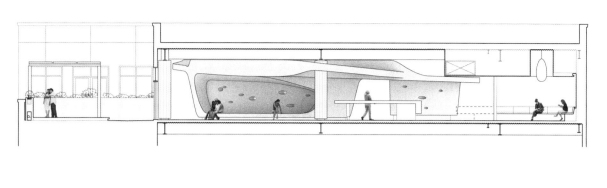
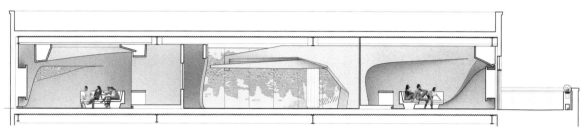
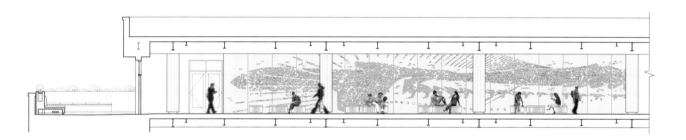
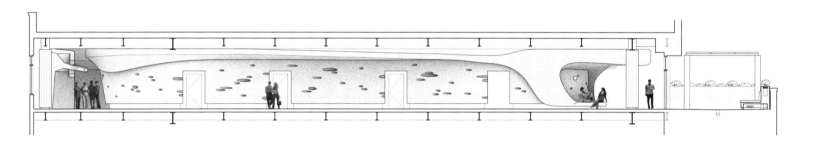

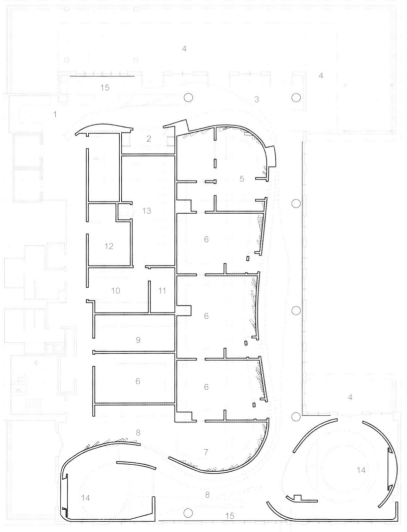

Plan:
1. Lobby
2. Kitchen
3. Common area
4. Terrace
5. Office
6. Project room
7. Conference room
8. Open office
9. Edit room
10. Tape op room
11. Scan
12. Film/tape vault
13. Machine room
14. Garding room
15. Aluminum laser-cut panels

平面图：
1. 大厅
2. 厨房
3. 公共区
4. 平台
5. 办公室
6. 项目室
7. 会议室
8. 开放式办公区
9. 编辑室
10. 磁带处理室
11. 扫描室
12. 胶片/磁带保管室
13. 机房
14. 保卫室
15. 激光切割铝板

Registrar Recorder County Clerk Elections Operations Center

登记员／县书记官选举运营中心

Jury Comments:
Graphic strength at its best – this design dramatically transforms a mundane warehouse into an energetic, highly functional, and aesthetically pleasing place of work.
Working with a modest budget and minimal means, the designer turned this space into a celebration through a skillful use of color, scale and graphics. The collaborative effort between the architect, client and artist is very successful and commendable.

评委评语：
图形的力量得到了最大的发挥——设计将平淡无奇的仓库改造成为一个充满活力、实用性强而美观舒适的办公空间。
在紧缩的预算和简单的设计方法的前提下，设计师为空间注入了色彩、规模和图形。建筑师、委托人和艺术家的合作成功而值得钦佩。

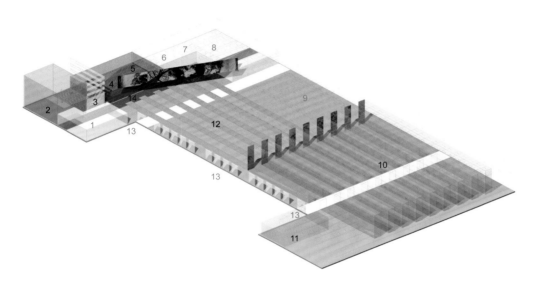

Diagram:
1. Call Center
2. Office
3. Café
4. Restrooms
5. Fire Rated Vault
6. Machine Shop
7. Ballot Inspection Room
8. Ballot Assembly Room
9. Operations Area
10. High Pile Storage Area
11. Equipment Storage Area
12. Staging Area
13. Loading Docks
14. Red "Hall"

图表：
1. 客户服务中心
2. 办公室
3. 咖啡厅
4. 餐厅
5. 耐火仓库
6. 机械车间
7. 投票检查室
8. 投票分配室
9. 操作区
10. 高架储藏区
11. 设备储藏区
12. 准备区
13. 装卸码头
14. 红厅

Notes of Interest

Registrar Recorder County Clerk Elections Operations Center is a 110,000-square-foot office and warehouse facility that organizes, distributes, collects and processes all of the voting materials of Los Angeles' 5,000 voting precincts for every election. The facility is housed in an existing tilt-up concrete warehouse, a structure of overwhelming size that houses all worker office space, voting pallet storage, digital voting units and personal records.

Designed and built-out in approximately 11 months, the project was brought in – all amenities included – under budget and on schedule. A robust palette of design tools were used to create a place of economy, utility and delight. The architect envisioned large-scale mega-banners that could achieve significant architectural impact. This new technology affordably provides scale and intimacy to this huge warehouse. In a matter of creative funding, the architect/owner team worked with the County Arts Program to commission a local artist. Color was used strategically on vertical and horizontal surfaces with paint and fabric, as well as through mega-banner technology. This use of color and imagery energizes the entire warehouse.

It was the intimate collaboration of architect, owner, user, builder and artist that allowed the success this project achieves. The result is a place of utility and delight, honoring an important institution of democracy as well as the citizens of Los Angeles County.

Owner: Los Angeles County
所有人：洛杉矶县

Architect / 建筑师
Lehrer Architects
莱勒建筑事务所

Location / 项目地点
Santa Fe Springs, California
加利福尼亚州,圣达菲斯普林斯

Photo Credit / 图片版权
© Michael B. Lehrer
迈克尔·B·莱勒

项目特色

登记员／县书记官选举运营中心总面积约10,220平方米的办公和仓库设施,它负责组织、分配、选择并处理洛杉矶县每次选举中5,000个选区的选举材料。中心坐落在一座上翘的混凝土结构仓库中,这个巨大结构里设置着所有办公空间、选举资料储藏室、数字投票装置和个人记录。

项目的设计和建造时间约11个月,出色地完成了预算和预期任务。粗犷型设计工具的选择打造了一个经济、实用而令人愉快的空间。建筑师利用跨度巨大的横幅来达到富有影响力的建筑效果。这个全新的技术为大型仓库划分出规模比例和私密空间。由于创意资金有限,建筑师和所有人与县艺术项目协会共同委托了一位本地艺术家进行设计。垂直和水平方向的平面上被色彩所覆盖,巨型横幅也为空间增色不少。色彩和图像的运用令整个仓库活力十足。

建筑师、所有人、使用者、建设者和艺术家的通力合作让项目取得了巨大成功。它兼具实用性和舒适性,突出了这个民主结构的重要性,也向洛杉矶县公民表示了尊重。

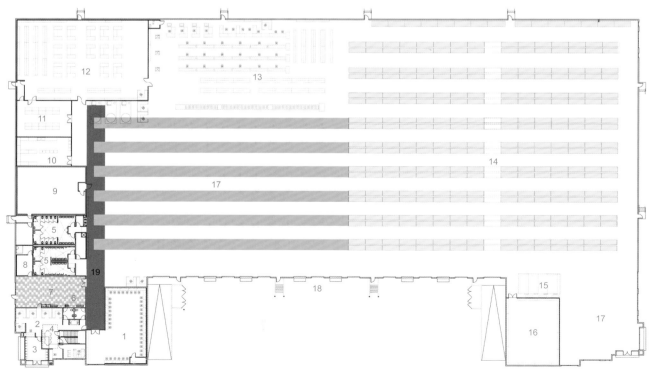

1. Call Center
2. Office
3. Waiting Area
4. Conference Area
5. Restroom
6. Kitchen
7. Café
8. Server Room
9. Fire Rated Vault
10. Machine Shop
11. Ballot Inspection Room
12. Ballot Assembly Room
13. Operations Area
14. High Pile Storage Area
15. Fork Lift Recharge Stations
16. Equipment Storage Area
17. Staging Area
18. Loading Docks
19. Red "Hall"

1. 客户服务中心　11. 投票检查室
2. 办公室　　　　12. 投票分配室
3. 等候区　　　　13. 操作区
4. 会议区　　　　14. 高架储藏区
5. 洗手间　　　　15. 叉车充电站
6. 餐厅　　　　　16. 设备储藏区
7. 咖啡厅　　　　17. 准备区
8. 服务器机房　　18. 装卸码头
9. 耐火仓库　　　19. 红厅
10. 机械车间

The Academy of Music
音乐学院剧院

Jury Comments:
Beautiful execution of historic preservation. The fact that this entry was so well documented and expertly executed, down to every faithful detail, makes us proud to revere the past and keep it ever present.
A thoughtful, meticulous restoration in which technical improvements are ingeniously concealed, and lighting is carefully placed to draw attention to the proportions, color and detailing that reawakens the space's unique character.
Sensitive and masterful.

评委评语:
出色的历史保护工程。设计的归档和执行都十分专业,进入到每个详实的细节,让我们骄傲地回顾过去、展望未来。
这是一次精心而细致的修复工作。技术设施被巧妙地隐藏起来,灯光设施布置得当,从比例、色彩到细部设计,无一不重现了空间的独特魅力。
敏感而出众。

Notes of Interest

The Academy of Music is the oldest continuously operating concert hall in the United States. The first performance occurred on January 26, 1857, but by 2007, years of continuous use had taken its toll on the Academy's Ballroom. The 40' x 80' Ballroom is spatially unchanged from 1857, but unfortunately many other historic features of the room were changed over time.

Through the generosity of Lee Annenberg and the Academy's ongoing restoration fundraising efforts, the Academy was in a position to undo 152 years of alterations. Hundreds of hours of design research went into reconstructing the original design intent for the room, using the Academy's substantial archives, as well as resources from The Philadelphia Historical Commission, The Pennsylvania Historical Society, and The Athenaeum of Philadelphia.

Lengthy efforts to repair the original room went into play, including the reintroduction of the chandeliers and gas light fixtures; the restoration of the glass windows, which had since been walled over and covered with mirrors; months of work reestablishing the paint scheme; and reparation of all water damage, hand in hand with steps taken to prevent future damage.

Engineer: Keast & Hood, PHY Engineers Inc.
General Contractor: L.F. Driscoll Company
Lighting: Horton Lees Brogden Lighting Design
Owner: The Philadelphia Orchestra
工程师:基斯特&胡德、PHY工程公司
总承包商:L.F.德里斯科尔公司
景观建筑师:霍顿・李斯・布罗戈登照明设计
所有人:费城交响乐团

Architect / 建筑师	**Location** / 项目地点	**Photo Credit** / 图片版权
KlingStubbins 克林斯德宾斯	Philadelphia, Pennsylvania 宾夕法尼亚州，费城	© Tom Crane Photography 汤姆·克莱恩摄影

项目特色

音乐学院剧院是美国最古老的持续运营中的音乐厅。它的第一场演出始于1857年1月26日。但是到了2007年,多年以来的持续运营让舞厅年久失修。自1857年以来,舞厅的空间一直保持原样,但是其他历史特征却已经时间变迁而改变了。

剧院得到李·安纳伯格和学院的维修募款支持,得到机会可将剧院恢复到152年前的原貌。数百小时的设计研究工作通过剧院档案和费城历史委员会、宾夕法尼亚历史协会和费城图书馆的帮助重现了剧院的原始设计。

大量的工作让原始剧院重新工作。其中包括:重新引入吊灯和煤气灯,修复被墙面堵住和被镜子挡住的玻璃窗,重新上漆以及修理漏水损坏的部分。

Ballroom: 舞厅平面图
1. Roof below — 1. 屋顶下方
2. Up — 2. 上
3. Dn — 3. 下
4. South grand stair — 4. 南侧大楼梯
5. Storage — 5. 仓库
6. Sound & light locks — 6. 声控&光控锁
7. Ambulatory — 7. 回廊
8. Exterior balcony — 8. 露天阳台
9. North grand stair — 9. 北侧大楼梯
10. Ballroom — 10. 舞厅

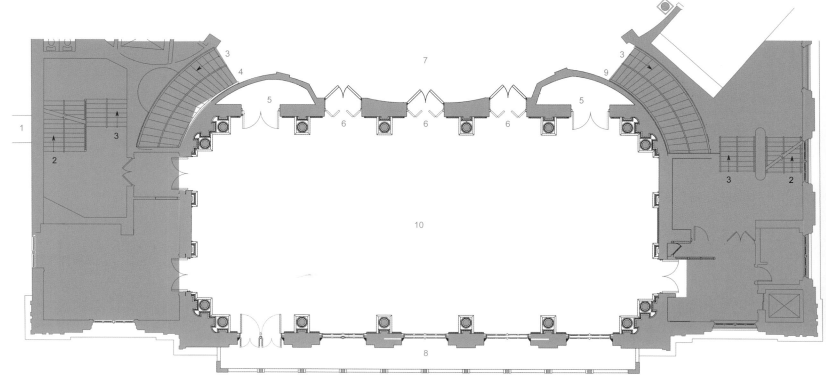

The Power House, Restoration/Renovation
电力站修复翻新

Jury Comments:
An interesting dialogue between present and past...The project maintains, preserves and reestablishes the integrity of the existing, historic building while creating modern, attractive, and energy efficient interior spaces that accommodate user needs.
A balanced and pleasing chiaroscuro effect is produced between new materials reflecting light and the existing materiality of the project that absorbs light.

评委评语：
项目在现代和过去之间形成了有趣的对话。项目维护、保护并重建了原有历史建筑物的完整性，创造出现代、富有魅力而节能的室内空间。

新旧材料之间形成了平衡而令人愉悦的明暗效果：新材料反射光线，旧材料吸收光线。

Engineer: Ruofei Sun, Ph.D., PE
General Contractor: R.G. Ross
Owner: Cannon Design
工程师：罗菲·孙（博士）
总承包商：R·G·罗斯
所有人：加农设计

Notes of Interest

These offices for the design architects' firm occupy a long-abandoned power house, constructed in 1928, part of the Municipal Service Building complex that still occupies an entire block of downtown St. Louis. The Power House component of the complex, designated as a landmark by the National Historic Register, had confounded developers over the years who struggled with its tall volume but relatively small footprint.

The design challenge was to accommodate 32,000 square-feet of office, conference and support space for approximately 120 employees in a building with 19,000 square-feet of floor area, but over 400,000 cubic-feet of volume. New floors needed to be added within the building's massive volume to accommodate the firm's program, but somehow this introduction of multiple stories had to avoid compromising the spatial integrity of its interior space, and most importantly, the 26-foot-tall, arched, revival-style windows. At the same time, the design needed to preserve each individual's connection to the exterior from the workspace. Lastly, it needed to meet and reflect the firm's desire for a new way of working: one that was intuitive, flexible and open. The second and third levels were added to massive existing steel plate columns – no vertical structure was added. Crisp, modern workspace is juxtaposed against rusted columns and glazed brick. The new floors are held away from the north and east elevations, which contain the dramatic Romanesque windows facing out to the city. The windows afford a significant amount of daylight and views to the surrounding neighborhood. The gallery space is programmed as an event space and a way to engage the community as never before possible.

项目特色

这座电力站建于1928年，是圣达菲市政服务大楼的一部分。项目将其改造成为设计建筑公司的办公室。电力站建筑已经被列入美国国家历史名迹之中，它细高狭窄的造型一直困扰着开发商们。

设计要求在这个占地面积1,765平方米的建筑内为120名员工提供近3,000平方米的办公、会议和相关辅助空间。项目必须在建筑内新增楼层，但是新增楼层又不能破坏整个室内空间的完整性，尤其是近8米高的文艺复兴风格拱形玻璃窗。同时，设计需要保留办公区与户外的连接性。设计满足并反映了公司对办公环境的需求：直观、灵活而开放。

建筑的二三层被添加在原有结构的大型柱子上——设计没有增添垂直结构。清新、现代的办公空间与锈迹斑斑的柱子和釉面瓷砖形成了鲜明对比。新增的楼层避开了建筑的东、北两个立面（罗马式窗口的所在地）。窗户为室内提供了足够的自然光线和周边的景色。走廊被打造成一个活动空间，前所未有地与社区联系了起来。

Architect / 建筑师	Location / 项目地点	Photo Credit / 图片版权
Cannon Design 加农设计	St. Louis, Missouri 密苏里州，圣路易斯	© Gayle Babcock/Architectural Imageworks, LLC, 盖尔·巴布科克/建筑图像公司

Second Floor Plan　二层平面图

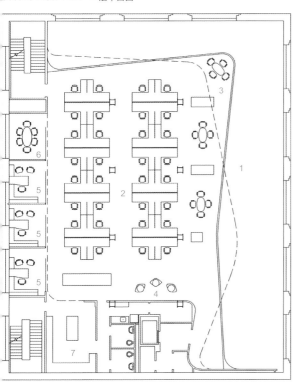

First Floor Plan　一层平面图

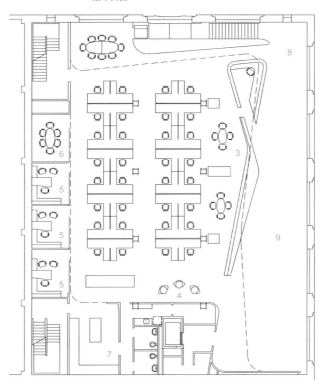

Plans:
1. Reception
2. Gallery
3. Open Office
4. Open Team Space
5. Crit Space
6. Office
7. Conference
8. Work Room
9. Open to Gallery Below

平面图（左二图）：
1. 前台
2. 走廊
3. 开放办公区
4. 开放团队空间
5. 讨论区
6. 办公室
7. 会议室
8. 工作室
9. 通往下方走廊

Vancouver Convention Center West
温哥华会展中心西区

Jury Comments:
An amazingly inviting, warm public circulation is achieved through the use of daylighting, building landforms, and local materials that both reference the area's industry and provide richly detailed surfaces.
Unlike the typical convention center, this public space has a well chosen vocabulary of materials and spatial proportions that commits to the connection between interior and exterior, bringing the outdoors in to its interior spaces in a compelling and eloquent manner.

评委评语：
照明、建筑地形和本地材料的运用打造出一个诱人而温馨的公共空间。本地材料体现了当地的工业背景，也提供了丰富的材质表面。
与典型的会展中心不同，项目的公共空间对材料和空间按比例的选择尤为出色。它有效地连接了室内外空间，采用引人注目的方式将室外空间引入室内。

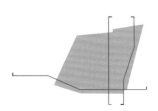

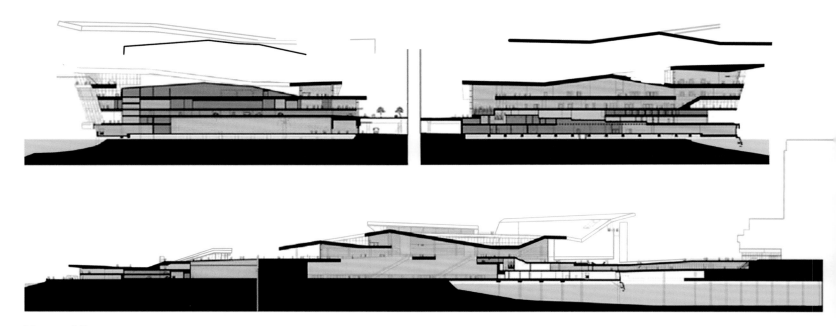

Notes of Interest

The design of the new Vancouver Convention Center West presented an opportunity to fully engage the urban ecosystem at the intersection of a vibrant downtown core and one of the most spectacular natural ecosystems in North America. Certified LEED® Canada Platinum, the project weaves together architecture, interior architecture, and urban design in a unified whole that functions literally as a living part of both the city and the harbor.

Addressing the human environment, the architectural approach creates a community experience that is simultaneously a building, an urban place, a park and an ecosystem. The convention center program emphasizes spaces for both public and private events, gatherings and circulation.

The primary interior expression is the use of naturally occurring materials indigenous to British Columbia, with extensive use of sustainably harvested Douglas fir. The ballroom and meeting room programs that form the core of the building's interior mass are enclosed by a wood cladding system that simulates the texture and directionality of a stack of lumber. Wood ceiling slats oriented in long, dramatic parallel lines combine with the orthogonal massing of the interior spaces to create contrast against the organic geometry of the roof and exterior shell. The strong wood expression takes on an arresting public presence at night as the building glows through its transparent skin.

Engineer: Glotman Simpson Consulting Engineers, EarthTech (Canada) Inc., Stantec Consulting, Schenke/Bawol Engineering Ltd., Sandwell Engineering Inc.
General Contractor: PCL Construction Enterprises
Owner: BC Pavilion Corporation (PavCo)
工程师：格罗特曼·辛普森工程咨询公司、大地技术公司（加拿大）、斯坦泰克咨询公司、施恩克/巴沃尔工程公司、桑德维尔工程公司
总承包商：PCL 建筑工程公司
所有人：BC 场馆公司

Architect / 建筑师
LMN + DA/MCM
LMN + DA/MCM

Location / 项目地点
Vancouver, Canada
加拿大，温哥华

Photo Credit / 图片版权
© Nic Lehoux
尼克·勒胡克斯

项目特色

温哥华会展中心西区的设计将城市生态系统与活跃的城市中心区以及北美最壮丽自然生态系统所结合。项目获得了加拿大绿色建筑白金认证，集建筑、室内设计和城市设计于一身，为城市和港口打造了一个鲜活的功能空间。

建筑设计以人为本，打造了一个兼具建筑、城市空间、公园和生态系统价值的整体区域。会展中心项目为公共和私人活动、集会、交通提供了空间。

项目的室内设计采用了不列颠哥伦比亚的本地自然材料，大量运用了具有可持续特性的花旗松。宴会厅和会议室是建筑内部的核心结构，采用了仿造木块材质和纹理的木质包层系统。木质天花板条以其细长的平行结构与室内空间的正交直线形成了鲜明的对比。夜晚，当建筑的透明表皮发出光芒，富有表现力的木质结构变得尤为醒目。

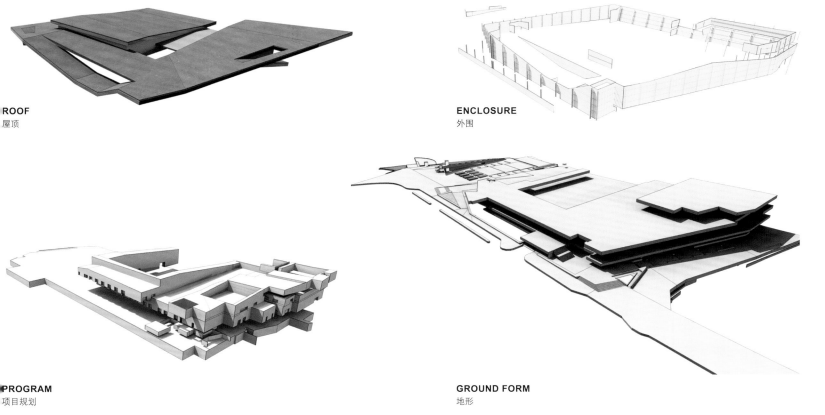

ROOF
屋顶

ENCLOSURE
外围

PROGRAM
项目规划

GROUND FORM
地形

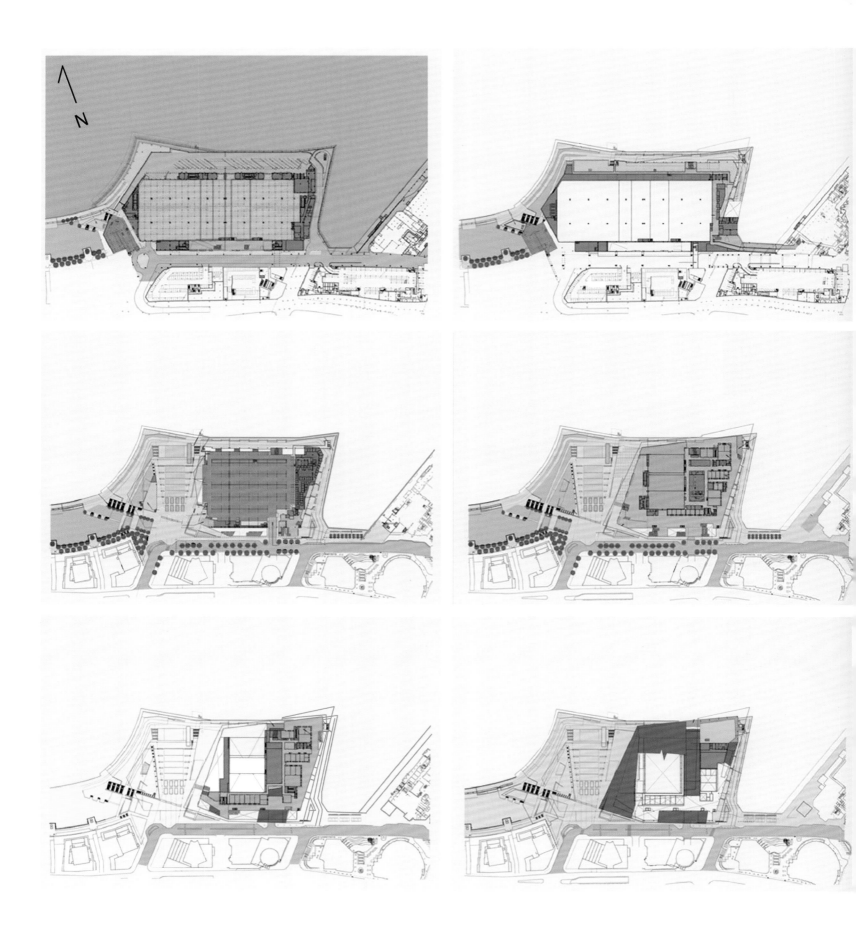

Washington Square Park Dental
华盛顿广场公园牙科诊所

Jury Comments:
Clever details and ideas are all over this skinny-budget project. It delightfully shows what you can do with a small space through a design that is visually very interesting and highly functional.
The use of serene glowing material, soothing color, simple planes of form, and a clear, open, yet structured, plan calms the nerves and sets a stage like no other dentist office.

评委评语：
巧妙的细部设计和理念在项目中随处可见。项目展示了小空间是如何通过设计变得吸引人而具有使用性的。平和的发光材料、治愈的色彩、简单的平面造型以及明晰开放的布局缓解了患者紧张的神经，营造出与众不同的牙医诊所。

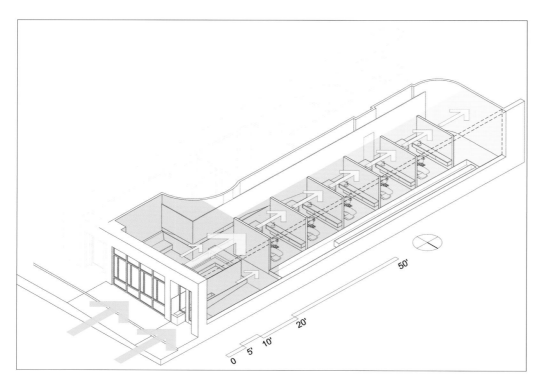

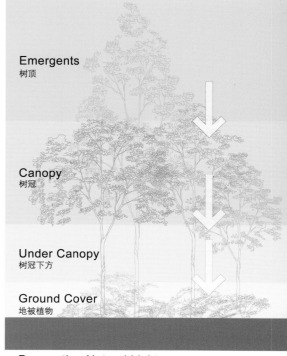

Notes of Interest

For this dentistry office located on the west edge of Washington Square Park in San Francisco, the main design goals involved maximizing the impression of space and length in the 1900-square-foot building and seamlessly bringing in elements of light and aesthetics from the exterior park all the way throughout the space.

Natural light floods into the space from the front windows all the way through to the back of the office space, despite the office's linear composition that includes five private patient operatory rooms. This is accomplished through the use of "floss" walls.

To extend the spatial depth of the relatively small space and establish a connection to the outside environment, aesthetic elements of the park were drawn into the interior. Low-irrigation interior landscaping is arranged in a beautiful steel perimeter separating polished river rocks from recycled glass. The long, interior entry ramp is framed by this linear garden, and it scontinues all the way fark the exterior of the building down the length of the space to create a series of individual gardens for each of the patient operatories. This view down the entire length of the office is immediately available to the patient upon entrance.

Engineer: Julia Y. Chen Design, Inc., Acies Engineering
General Contractor: Norcal Construction Management Services
Lighting: John Brubaker Architectural Lighting Consultants
Owner: Washington Square Park Dental
工程师：茱莉亚·Y·陈设计公司、艾西斯工程公司
总承包商：北加州建筑管理服务公司
照明设计：约翰·布鲁巴克尔建筑照明咨询公司
所有人：华盛顿广场公园牙科诊所

Architect / 建筑师
Montalba Architects, Inc.
蒙塔尔巴建筑公司

Location / 项目地点
San Francisco, California
加利福尼亚州，旧金山

Photo Credit / 图片版权
© Mitch Tobias
米奇·托拜厄斯

项目特色

这个牙科诊所坐落在旧金山华盛顿广场的西侧，主要设计目标是最大化空间表现力，为室内引入光线和公园的美景。

尽管诊所的五个私人治疗室呈线型排开，自然光线透过前窗洒入室内空间，直达后部的办公空间。"丝棉"墙壁的运用实现了这一效果。

为了扩展空间深度、建立与户外环境之间的联系，公园的美学元素被引入了室内。低灌溉型室内景观被设置在漂亮的钢槽里，旁边点缀着抛光的河流岩石。这个线型花园围起了长长的室内入口坡道，坡道一直延伸到建筑室外，为每个私人治疗室都提供了独立的花园。患者一进门就能欣赏到这个独特的景观。

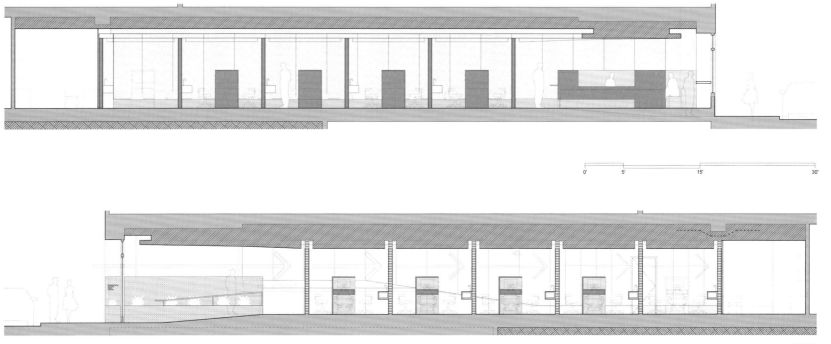

Sustainable Strategies – Exploded Axonometric:
1. High albedo folded celing plane
2. Translucent acrylic panels
3. Enkadrain mats
4. Steel stud construction
5. Fsc certidied wood
6. Recycled hot-rolled steel panels
7. Transparent floss wall assembly
8. Refurbished detail chairs
9. No voc paint
10. Low irrigation interior landscape with recycled glass
11. Carpet tiles from renewable and recycled sources
12. Perforated sound absorbing wall using fsc certified wood

可持续策略——分解轴测图：
1. 高反射率折叠天花板
2. 半透明亚克力板
3. 排水垫
4. 钢龙骨结构
5. 经过美国森林管理局认证的木材
6. 回收热轧钢板
7. 半透明"丝棉"墙装配
8. 翻新细部座椅
9. 无挥发涂料
10. 低灌溉室内景观，配有回收玻璃
11. 可再生和回收来源的块式地毯
12. 穿孔隔音墙板，采用了经过美国森林管理局认证的木材

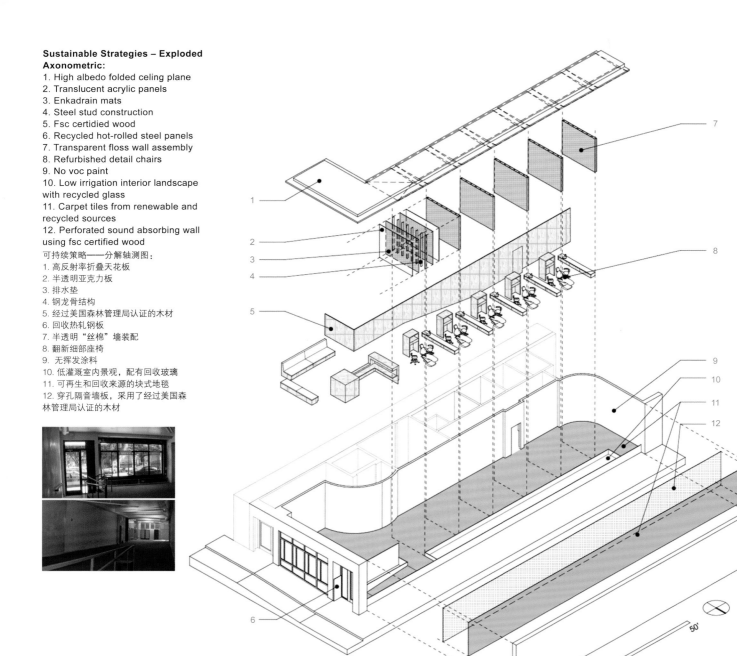

Floor Plan (Below):
1. Entry
2. Reception
3. Waiting area
4. Built-in seating
5. Custom perforated acoustic wall
6. Operatory
7. Custom translucent floss wall
8. Built-in storage
9. Landscape
10. Office

平面图（下图）：
1. 入口
2. 前台
3. 等候区
4. 嵌入式座椅
5. 定制的预制隔音墙
6. 手术室
7. 定制的半透明"丝棉"墙
8. 嵌入式储藏间
9. 景观
10. 办公室

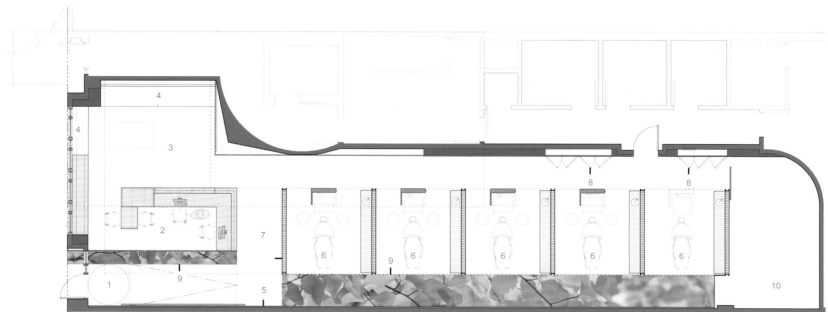

Transverse Building Section Through Typical Patient Operatory:
1. Fixed fsc wood paneling
2. White acrylic lens
   Continuous flour. Light fixture
   1/4" Thick hot-rolled steel panel
3. Recycled carpet tile
   Existing slab on grad or wood
   Framing (depending on location)
4. Opening for television in framing
5. 1/4" Frosted acrylic panels
   Nylon enkadrain mat
   Mc6x12 steel channel
6. Perforated acoustical panel
   3/4" Air gap
   Recycled carpet tile
   Existing demising wall
7. Interior landscape planter

标准手术室的横切面：
1. 固定的美国森林管理局认证木镶板
2. 白色亚克力透镜
   连续的灯具
   1/4''厚热轧钢板
3. 回收材料制成的块状地毯
   原有的石膏板或木板
   框架结构（根据地点的不同而不同）
4. 电视背景框

5. 1/4''磨砂亚克力板
   尼龙排水垫
   MC6X12钢槽
6. 穿孔隔音板
   3/4''气隙
   回收材料制成的块状地毯
   原有的活动墙
7. 室内景观植物

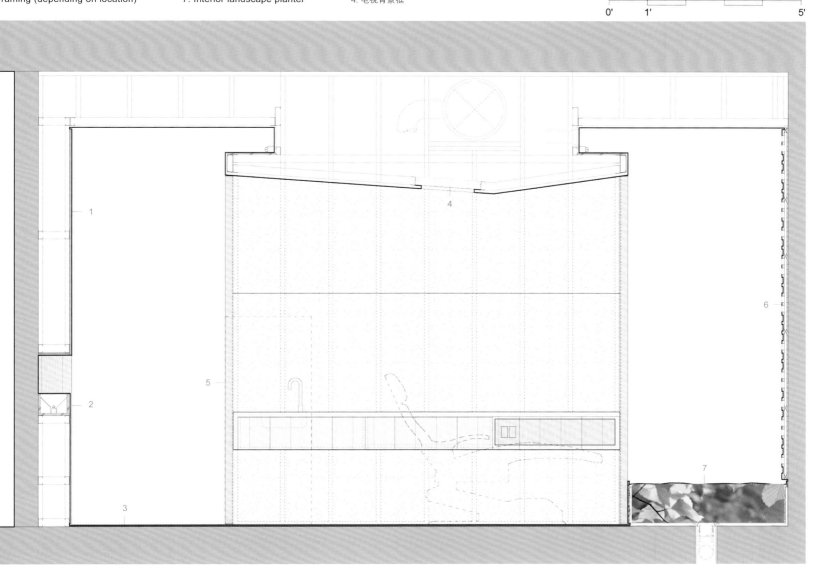

John E. Jaqua Center for Student Athletes
约翰·E·雅克大学生运动员中心

Jury Comments:
Beautifully detailed throughout...Remarkable introduction of color and management of natural light. The atmosphere is comfortable yet stimulating, with furnishings that are highly functional yet inviting and uniquely designed.
Athletics and Education are purely stated here. This facility's integration of diverse and defined spaces, of glass and daylight for engaging learning, of wood for warmth and strength, and with effective graphics celebrating its graduates' post-graduate achievements creates a design whose goals are transparent.

评委评语:
优美的细部贯穿项目始终。色彩和自然光线的处理非凡出众。整体氛围舒适而富有启发性，所有设施都既实用又令人愉悦。
体育和教育在此获得完美的表述。该中心结合了形形色色的空间元素：玻璃和日光让人专注学习，木材提供了温暖和力量感，图案图形展示了校友的成就。设计目标十分明确。

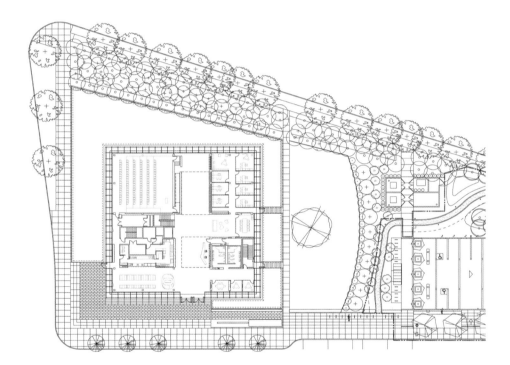

Notes of Interest

The John E. Jaqua Center for Student Athletes at the University of Oregon is a bright space that emphasizes the value of academic and professional achievements to its athletic audience through design. This new 40,000-square-foot, state-of-the-art, academic learning center accommodates the NCAA-mandated academic services for the tutoring of 520 student athletes.

The challenging project site's visual prominence led to the building being designed with four public façades and no "back door", which has reinforced its importance within the landscape as an iconic element.

The notion of a fertile, natural environment to invigorate and inspire learning was the premise on which the design concept was based. The glass structure rests on a "table of water". A "double wall" façade addresses acoustic isolation, thermal insulation, and control of available daylight within the building. A prismatic, vertical stainless steel screen within this façade provides shading, thermal comfort, and ability for heat harvesting as well as visual privacy for the inhabitants. The glazed façade and interior spaces are composed on a rigorous module to achieve an uninterrupted visual connection between internal rooms and the larger garden beyond.

Also present in the project is the university's signature yellow and "O", as well as various achievement-centered art installations, all of which help the facility to serve as a pantheon of student athletic achievements.

Engineer: KPFF Consulting Engineers, Inc., Interface Engineering, Arup, Harper Houf Peterson Righellis, Inc.
General Contractor: Hoffman Construction Company
Owner: The University of Oregon
工程师：KPFF工程咨询公司、界面工程公司、阿拉普公司、哈珀·霍夫·彼得森·里格里斯公司
总承包商：霍夫曼建筑公司
所有人：俄勒冈大学

Architect / 建筑师
ZGF Architects LLP
ZGF建筑事务所

Location / 项目地点
Eugene, Oregon
俄勒冈州，尤金

Photo Credit / 图片版权
© Basil Childers
巴兹尔·切尔德斯

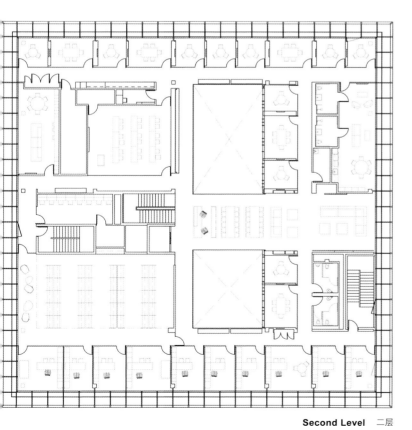

Second Level 二层

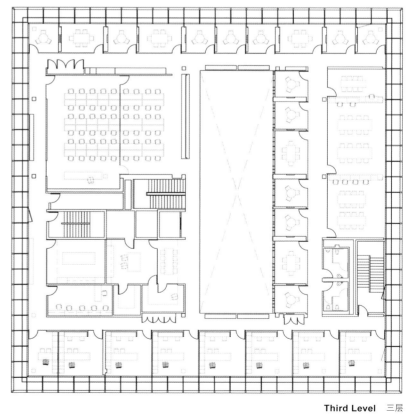

Third Level 三层

项目特色

俄勒冈大学约翰·E·雅克大学生运动员中心宽敞明亮,旨在通过设计突出大学生运动员的学术和专业成就。这个先进的学术中心为520名大学生运动员提供美国大学生协会所规定的学术服务。

项目场地所面临的挑战是建筑奖拥有四个公共外立面,并且没有后门,以突出其在景观中的重要性和地标作用。

富饶的自然环境将鼓舞并激发人的学习热情,这是设计概念的前提。玻璃结构建筑坐落在一个水台之上。双层墙面系统保证了建筑的隔音、隔热和光线控制。外立面内的棱柱形垂直不锈钢遮阳板起到了遮阳和热收集作用,同时也保证了内部人员的隐私。玻璃外立面和室内空间在楼内房间和广阔的花园之间形成了不间断的视觉联系。

项目还呈现了大学的标志"黄色的O"和各式各样的装置艺术,使中心成为了大学生运动员的圣殿。

Beijing CBD East Expansion
北京中央商务区东扩规划

Jury Comments:
A compelling vision for an expanded dense city core that is environmentally sound, with intertwined linear parks creating a variety of pedestrian opportunities, and believably defining and linking multiple districts into a livable city.
In the midst of great density, Beijing has found a sensitive way to incorporate environmental sustainability into its expanded business district. This project shows how planning for a cohesive central business district is the best strategy for truly sustainable development.

评委评语:
这是一个具有说服力的城市中心区扩张整体规划。带状公园的设计极具环保价值,提供了多种类型的步行空间,同时也将各个街区连接在一起,成为宜居型城市。
作为建筑密集度较高的城市,北京找到了一个能够将环保可持续开发纳入商务区扩张的解决方案。
项目展示了如何规划一个综合中央商务区,在可持续开发方面尤为出色。

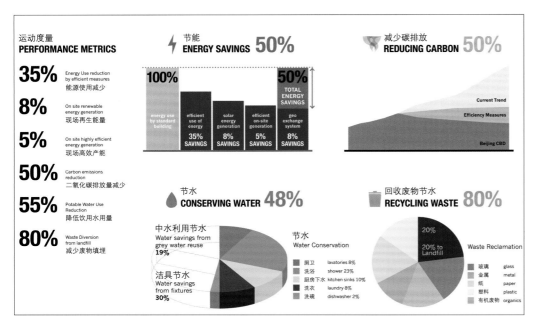
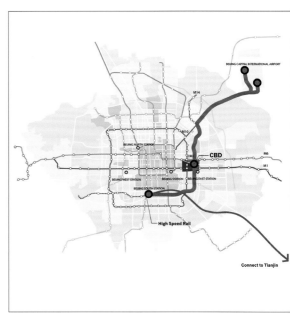

Notes of Interest

Located in the heart of Beijing, the Central Business District (CBD) has emerged over the past decade as China's primary global business address and is now poised for an eastward expansion that will almost double its size. The CBD Eastern Expansion Plan defines opportunities for the growth of commerce, industry, culture and the arts by establishing a flexible framework for growth and an environmentally sustainable approach to 21st Century city design.

Accommodating up to 7 million square meters of new development over a 3-square-kilometer site, the plan calls for a restored commitment to public open space and a heightened focus on connectivity and mobility through advanced public transportation systems. A district-wide intelligent infrastructure system, composed of integrated utilities and controlled by smart technology, enables the CBD to function at optimum efficiencies and creates a model for large-scale, low-carbon, urban development.

With public open space largely absent in the existing CBD, or at best scattered and disconnected, the plan calls for signature parks as the center of new urban districts. Green boulevards connect the parks to provide a comprehensive open space network.

Landscape Architect: The Office of James Burnett
Owner: Beijing CBD Administrative Committee
景观建筑师: 詹姆斯·伯内特工作室
所有人: 北京中央商务区行政委员会

项目特色

中央商务区位于北京城的中心,形成于21世纪初,是中国主要的全球商务区,现在正在寻求向东扩张。中央商务区东扩规划为商业、工业、文化和艺术的成长提供了机遇,将建立一个灵活的成长框架和21世纪城市规划可持续设计方案。项目规划涉及了在3平方公里内的700万平方米的新开发工程,要求打造充满活力的开放空间并通过升级版的公共交通系统建立更好的连通性和移动性。由综合公共设施和智能技术控制组成的跨区智能基础设施系统将保证中央商务区的最佳效率,打造大规模低碳城市开发工程的典范。
目前,中央商务区缺乏公共开阔空间(或者说仅有的公共空间也十分分散),规划要求在新建的城市街区内打造特色公园。而连接公园的绿色林荫大道将提供综合开放空间网络。

Architect / 建筑师	**Location** / 项目地点	**Photo Credit** / 图片版权
Skidmore, Owings & Merrill LLP SOM建筑事务所	Beijing, China 中国，北京	© Skidmore, Owings & Merrill LLP SOM建筑事务所

Chicago Central Area DeCarbonization Plan
芝加哥中心区脱碳规划

Jury Comments:
An innovative way of looking at long existing urban fabrics that uses common sense approaches to creatively solve common problems.
Kudos to Chicago for being one of the leading U.S. cities working on climate action.
This ambitious proposal shows how a dense urban area can effectively continue to reduce greenhouse gas emissions and lessen the carbon footprint of its business district.

评委评语:
项目创新地审视了现有的城市网络,利用常规方法创新地解决了普遍问题。
芝加哥堪称美国气候行动的领军城市之一。
这个雄心勃勃的方案展示了如何让密集的城市区域有效并持续地减少温室气体排放以及如何降低商业区域的碳排放量。

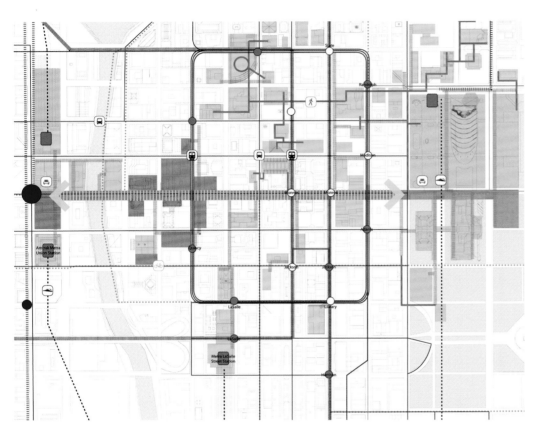

Notes of Interest
The Chicago Central Area DeCarbonization Plan is a comprehensive vision for helping the City of Chicago reach the goals of the Chicago Climate Action Plan and the 2030 Challenge in the downtown Loop. The project team developed a database (energy use, size, age, use, and estimated carbon footprint) of more than 550 buildings. The team used that database, tied to a 3-D model, to develop the DeCarbonization Plan, which interweaves energy engineering, architecture, and urban design.
In the DeCarbonization Plan's synergistic approach, eight key strategies work together with a parametric model. The first strategy, "Buildings", investigates how existing structures can be upgraded to improve efficiency, increase the value of aging building stock and tap into the potential to transfer excess energy back to the grid, all while offsetting the need for new construction. Second is "Urban Matrix", which envisions increasing the residential density of the Loop by enhancing amenities, adding schools and services and converting aging office buildings to residential.
Others strategies include "Smart Infrastructure", a look at how energy can be generated, stored, distributed and shared; "Mobility", assessing transit and connectivity; "Water", examining how this critical resource is used and Conserved; "Waste", examining systems for reducing, recycling and disposing of waste; "Community Engagement", involving citizens in the green agenda; and "Energy", an examination of existing and new energy sources.

Consultant: Bryan Cave LLP,
City of Chicago - Mayor's Green Team,
Environmental Systems Design, Inc.
Owner: City of Chicago Department of the Environment
顾问:布莱恩·卡夫公司、
芝加哥市市长绿色团队、
环境系统设计公司
所有人:芝加哥环境局

Architect / 建筑师
Adrian Smith + Gordon Gill Architecture
艾德里安·史密斯 + 戈登·吉尔建筑事务所

Location / 项目地点
Chicago, Illinois
伊利诺伊州，芝加哥

Photo Credit / 图片版权
© Adrian Smith + Gordon Gill Architecture
艾德里安·史密斯 + 戈登·吉尔建筑事务所

1 **Thompson Center** 汤普森中心：100W.Randolph 伦道夫街100W 1985 年 Helmut Jahn 建筑师 19 Stories 19层 Office/Retail 办公楼/零售 1,557,654GSF /平方英尺	2  **Daley Center** 戴利中心：50W.Washington 华盛顿街50W 1965 年 Murphy & Naess 建筑师 32 Stories 层 Office 办公楼 1,234,848GSF /平方英尺	3 **Daley Center** 市政厅：48N.Clark 克拉克街48N 1911年 Holabird & Roche建筑师 12 Stories 12层 Office 办公楼 1,352,946GSF /平方英尺	4 **Daley Center** 华盛顿大学：30E.Lake 雷克街30E 1962 年 11 Stories 11层 School 学校	
5 **Cook County Admin.** 库克郡行政大楼：69W.Washington 华盛顿大街69W 1965 年 Jacques Brownson 建筑师 35 Stories 35层 Office 办公楼 1,044,938GSF /平方英尺	6 **Fed. Plaza Post Office** 邮局联邦广场：230S.Dearborn 迪尔伯恩街230S 1975年 Mies van der Rohe 建筑师 1 Story 层 Post Office 邮局	7 **Federal Plaza Bldg.1** 联邦广场1号楼：219S.Dearborn 迪尔伯恩街219S 1969 年 Mies van der Rohe 建筑师 39 Stories 39层 Office办公楼 1,243,980GSF /平方英尺	8 **Federal Plaza Bldg.2** 联邦广场2号楼：230S.Dearborn 迪尔伯恩街230 1975 年 Mies van der Rohe 建筑师 45 Stories 45层 Office 办公楼 1,333,800GSF /平方英尺	
9 **Federal Plaza Bldg.** 联邦广场大楼：230S.LaSalle 拉萨尔街230S 1922/1960/1986 年 Graham, A, P& White 建筑师 14 Stories 14层 Office 办公楼 985,000GSF /平方英尺	10 **H.Washington Library** 华盛顿大学图书馆：400S.State 州立街400S 1991 年 Hammond, Beeby & Babka 建筑师 8 Stories 8层 Library 图书馆 756,640GSF /平方英尺			

项目特色

芝加哥中心区脱碳规划旨在帮助芝加哥市达到芝加哥气候行动方案和2030年市中心区挑战的目标。项目团队开发了一个涵盖550座建筑的数据库（能源利用、尺寸、年龄、使用方式以及预估碳排放量）。设计团队利用这个数据库与3D模型相结合，开发出融合了能源工程、建筑和城市规划的脱碳规划。

在脱碳计划的协同效应中，八个主要策略与一个参数模型共同作用。第一个策略"建筑"研究了如何升级现有建筑——提升效率、增加旧建筑价值并且挖掘转移多余能源重新进入网络的潜力，借以满足新建筑的能源需求。第二个策略是"城市矩阵"。它规划了通过增强设施建设、增加学校和服务设计以及进行旧楼改造来提升中心区住宅密度。

其他的策略包括"巧妙的基础设施"——如何产生、存储、分配和共享能源；"移动性"评估了运输和连通性；"水"检查了这个关键资源的使用和节约；"废物"检查了垃圾的减少、回收和处理系统；"社区参与"融入了市民在绿色议程中的参与；"能源"检查了现有能源和新能源的使用情况。

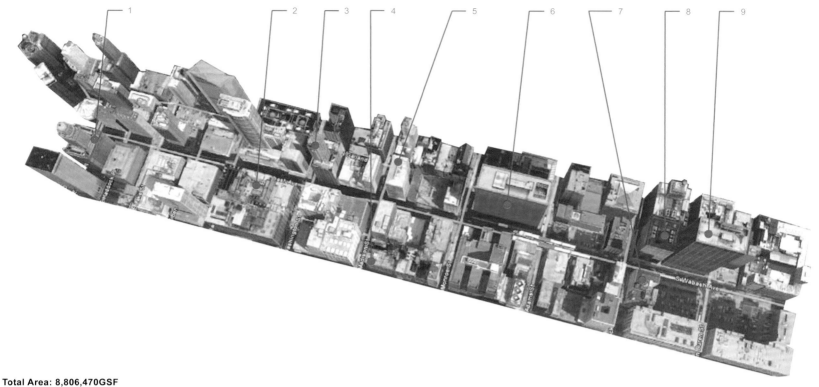

Total Area: 8,806,470GSF
总面积：818,148平方英尺

Jewelers Building
珠宝大厦：
35E..Wacker 瓦尔克街35E
1927 年
Thielbar & Frugard 建筑师
40 Stories 40层
Office 办公楼
966,720GSF/平方英尺

Marshall Fields
马歇尔商店：
122N.Wabash 沃巴什街35E
1914 年
Graham,Burnham&Co. 建筑师
12 Stories 12层
Department Store 百货公司
1,600,000GSF/平方英尺

Pittsfield Building
皮兹菲尔德大厦：
55E.Washington 华盛顿大街55E
1956-1927 年
Graham, Anderson,
Probst,and White 建筑师
38 Stories 38层
Office 办公楼
515,280GSF/平方英尺

Carson Pirie Scott:
斯科特百货大楼：
5S.Wabash 沃巴什街5S
1899-1961 年
O'Sullivan,Burnham 建筑师
12 Stories 12层
Office/Dept.Store 办公/百货
110,640GSF/平方英尺

Mallers Building
麦勒尔斯大厦：
5S.Wabash 沃巴什街5S
1910 年
Christian Eckstrom 建筑师
21 Stories 21层
Office 办公楼
352,296GSF/平方英尺

Mid Continental Plaza
中大陆广场：
55E.Monroe 门罗街55E
1972 年
Goettsch Parners Inc. 建筑师
49 Stories 49层
Office 办公楼
2,899,394GSF/平方英尺

DePaul College of Law
德保罗法学院：
25E.Jackson 杰克逊街35E
1917 年
Graham, Burnham & Co 建筑师
16 Stories 16层
School 学校
540,000GSF/平方英尺

Carson Pirie Scott:
CAN北广场：
55E.Jackon 杰克逊街55E
1962 年
Naess & Murohy 建筑师
24 Stories 24层
Office 办公楼
522,140GSF/平方英尺

Mallers Building
CAN广场大厦：
333S.Wabash 沃巴什街333S
1973 年
Graham, Anderson,
Probst, and White 建筑师
44 Stories 44层
Office 办公楼
1,300,000GSF/平方英尺

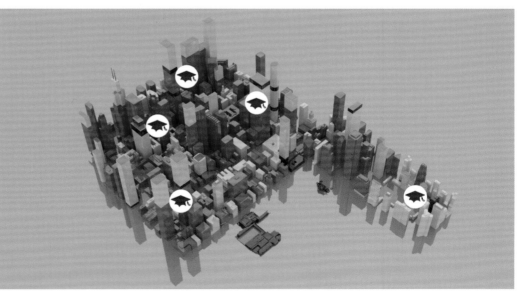

"Community | City: Between Building and Landscape Affordable Sustainable Infill for Smoketown, Kentucky"

"社区|城市：烟镇建筑与景观之间的可持续填充设计"

Jury Comments:
What a thorough analysis of the opportunity for intervention into a typical town that elevates the conversation regarding infill and landscape while respecting history and increasing density. Very applicable to many other places.
This project increases density while sensitively reinforcing the city's historic fabric and reactivates interstitial neighborhood spaces to produce a robust public realm.
The challenges facing Smoketown are found in small and large communities everywhere. Likewise, the plan for Smoketown, with its emphasis on infill development, sustainable landscaping, and sensitivity to cultural and historic components, is a model for communities near and far.

评委评语：
这是一次对城镇彻底改造的机会，在尊重历史和增加城市密度的前提下提升了城市和景观之间的对话。项目也适用于其他许多地区。
项目既增加了住宅密度又提升了历史城市结构。社区空间通过强有力的公共区域重新变得活跃起来。
烟镇所面临的挑战存在于随处可见的大小社区。同样的，烟镇规划以填充式开发、可持续景观和文化与历史元素为重点，是远近社区开发的典范。

Consultant: Architectural Artisans,
Clarksdale + Smoketown Infill,
K. Norman Berry Associates Architects LLP,
Louis and Henry Group
Project Research: University of Kentucky College of Design, University of Kentucky/University of Louisville Urban Design Center
Research Participants: Bates Memorial Baptist Church/Community Development Corporation,
HOPE VI Development,
The Housing Partnership,
The Kentucky Housing Corporation,
Louisville Metro Housing Authority,
Louisville Metro Parks,
Louisville Metro Planning,
The Olmsted Conservancy,
Smoketown / Shelby Park Residents,
Smoketown Presbyterian Community Center
Owner: City of Chicago Department of the Environment
顾问： 建筑工匠、克拉克斯戴尔+烟镇填充设计、K·诺曼·巴里建筑事务所、路易斯和亨利集团
项目研究： 肯塔基大学设计学院、肯塔基大学/路易斯维尔大学城市规划中心
研究人员： 贝茨纪念浸礼会教堂/社区开发公司、希望VI开发公司、住房合作公司、肯塔基住房公司、路易斯维尔地铁房屋署、路易斯维尔地铁公园、奥姆斯特德保护局、烟镇/谢尔比公园住宅、烟镇长老社区中心
所有人： 芝加哥环境局

Notes of Interest
This project remediates existing brownfields and re-activates a long-neglected connection among an historic African American residential neighborhood, an historic Olmsted park, and the Ohio Riverfront. Gaps are filled in an existing neighborhood fabric, increasing density while sensitively reinforcing its historic urban structure. Neighborhood spaces are re-activated to produce a robust public realm.
Full advantage is taken of the project's historic urban location, and its walking-distance proximity to neighborhood amenities. It encourages intense use of the residual spaces between structures as a catalyst, fostering to community identity and a renewed sense of place.

Architect /建筑师
Marilys R. Nepomechie Architect and
Marta Canaves Interior Design
马里莱斯·R·耐波米奇建筑事务所和玛尔塔·卡纳维斯室内设计

Location /项目地点
Smoketown, Kentucky
肯塔基州，烟镇

Photo Credit /图片版权
© Marta Canaves
玛尔塔·卡纳维斯

Compost 4"	堆肥 4"
Fertilizer 1"	化肥 1"
Straw 8"	稻草 8"
Fertilizer 1"	化肥 1"
Hay 4"	干草 4"
Newspaper Filler 1/4"	报纸填充 1/4"
Frame 8"-10"	框架 8"–10"

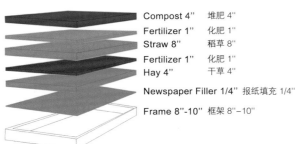

LEED ND | 绿色建筑策略

Smart Location and Linkages + Neighborhood Pattern and Design + Green Infrastructure and Buildings

巧妙的定位和连接+周边模式和设计+绿色基础设施和建筑

1. Vegetation — 1. 植物
2. Growing Medium — 2. 生长媒介
3. Drainage, Aeration, Water Storage and Root Barrier — 3. 排水、通风、储水和根部绝缘
4. Insulation — 4. 隔离层
5. Membrane protection and Root Barrier — 5. 薄膜保护和根部绝缘
6. Roofing Membrane — 6. 屋顶薄膜
7. Structural Support — 7. 结构支持

Urban Farming: Roof Terraces
城市框架：屋顶平台

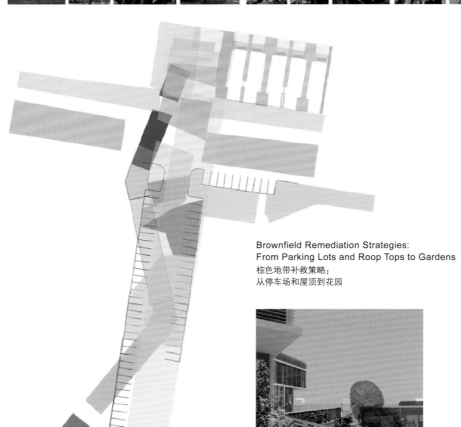

Brownfield Remediation Strategies:
From Parking Lots and Roop Tops to Gardens
棕色地带补救策略：
从停车场和屋顶到花园

Greening Urban Corridors | Re-Connecting to the Riverfront   绿色城市走廊｜重新与河岸相连

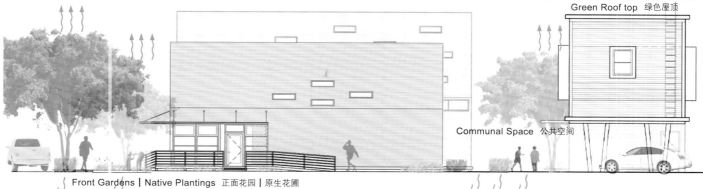

Green Roof top 绿色屋顶

Communal Space 公共空间

Front Gardens | Native Plantings 正面花园 | 原生花圃

Native Perennials: Little Bluestem; Lobelia; Sea Oats; Butterfly Weed; Copper; Yellow and Blue Iris; Sensitive Fern; Oak leaf Hydrangea
These help to soak up water and filter pollutants that enter with rain water.
多年生植物：小须芒草；半边莲；海燕麦；马利筋；铜草；
黄色和蓝色鸢尾；球子蕨；橡叶绣球花
这些原生植物吸收水分封并过滤雨水中的污染物。

LEED ND 绿色建筑策略

Smart location and Linkages + Neighborhood Pattern and Design + Green Infrastructure and Buildings
巧妙的定位和连接+周边模式和设计+绿色基础设施和建筑

Permeable Pavers and other Meaterials to:
- Contribute to LEED points
- Capture and treat the first flush
- Reduce runoff by 100% for low intensity storms
- ADA compliant

可渗透路面和其他材料能够：
- 有助于绿色建筑策略
- 捕捉和处理初次雨水冲刷
- 减少低强度风暴100%径流
- 适应农业开发

Terrace communal garden 平台公共花园

Green Roof top 绿色屋顶

Infiltration and filter through permeable materials and plants
可渗透材料和植物保证了渗透和过滤

(RE) BUILDING COMMUNITY: （重新）建设社区：
GREENING the IN-BETWEEN 在中间添加绿化

项目特色

项目修复了原有的棕色地带（污染区），并且将非裔美籍住宅区、奥姆斯特德公园和俄亥俄河联系了起来。原有社区网络之间的空缺得到了填充，既增加了住宅密度又提升了历史城市结构。社区空间通过强有力的公共区域重新变得活跃起来。

项目充分利用了自身的地理位置和便利的交通条件。它支持人们利用建筑之间的剩余空间作为催化剂，培养社区形象，为区域带来全新的感觉。

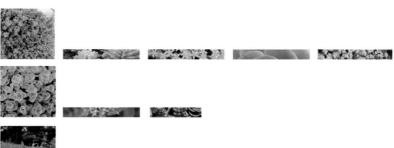

Community Roof Gardens and Terraces
社区屋顶花园和平台

Gowanus Canal Sponge Park
郭瓦纳斯运河海绵公园

Jury Comments:
The inspiring approach that can apply to many of our cities' neglected waterfront industrial sites containing a key element in urban transformation and water management.
An inspiring plan that can coexist with its industrial neighbors while redefining the word industrial and waterfront at the same time.
This project addresses a common problem in older cities with a striking way to create valuable urban open spaces as a byproduct.

评委评语：
这一极具启发性的设计能够被应用于城市中许多被我们遗忘的滨水工业区，包含了城市改革和水处理过程中的重点元素。
项目富有启发性，与现有工业区域共存，并且重新定义了工业和滨水区。
项目解决了旧城市中的普遍问题，同时也创造了极具价值的城市开放空间。

Funding: The Gowanus Canal Conservancy, The New England Interstate Water Pollution Control Commission The New York State Council on the Arts
Owner: City of New York, Department of Environmental Protection
资金来源： 新英格兰洲际水域保护控制委员会、纽约州艺术委员会
所有人： 纽约市环境保护局

Notes of Interest

The Gowanus Canal Sponge Park™ is a public open space system that slows, absorbs and filters surface water runoff with the goal of remediating contaminated water, activating the private canal waterfront, and revitalizing the neighborhood. The total proposed area for the Gowanus Canal Sponge Park™ system is 11.4 acres: 7.9 acres of esplanade and recreational open spaces, and 3.5 acres of remediation wetland basins.

The most unique feature of the park is its character as a working landscape: its ability to improve the environment of the canal over time while simultaneously supporting public engagement with the canal ecosystem. New York City has a combined sewer system. Rain that falls within the Gowanus watershed enters the storm drains and mixes with raw sewage in the sanitary sewer system. In a heavy rainfall, the combined sewage and storm water overflow directly into the Gowanus Canal.

The innovative Sponge Park™ plan proposes diverting surface water runoff into a water management system. The storm water management system activates the corridors leading from the adjacent neighborhoods to the park esplanade, while preventing further contamination Cof the canal. The parks incorporate vegetated landscape buffers to slow, percolate and filter the contaminated water, reducing the input of stormwater into the sewer system while integrating programmed urban outdoor space to create an accessible waterfront.

项目特色

郭瓦纳斯运河海绵公园是一个能够减缓、吸收并过滤地表水径流的公共空间系统，其目标是修复污染的河水、活跃私人运河滨水并复兴周边区域。郭瓦纳斯运河海绵公园的总规划面积为4.8公顷：3.2公顷为散步和休闲空间，1.6公顷为湿地修复空间。

公园最突出的特色为它是一项工作的景观：它的作用是提升运河环境并支持运河生态系统的公共处理方式。纽约市拥有一个合流制排放系统：郭瓦纳斯运河流域的雨水进入雨水道并且与生活污水系统中的废水混合。在暴雨时，混合的废水和雨水将直接流入郭瓦纳斯运河。

创新的海绵公园规划提出将地表水径流导入一个水处理系统。雨水处理系统将激活从附近区域到公园散步区的走道，同时也防止运河的进一步污染。公园的植被和景观缓冲区将减缓、浸透并过滤污染的河水，减少进入污水系统的雨水，同时也结合了城市户外空间，打造了可进入的滨水区域。

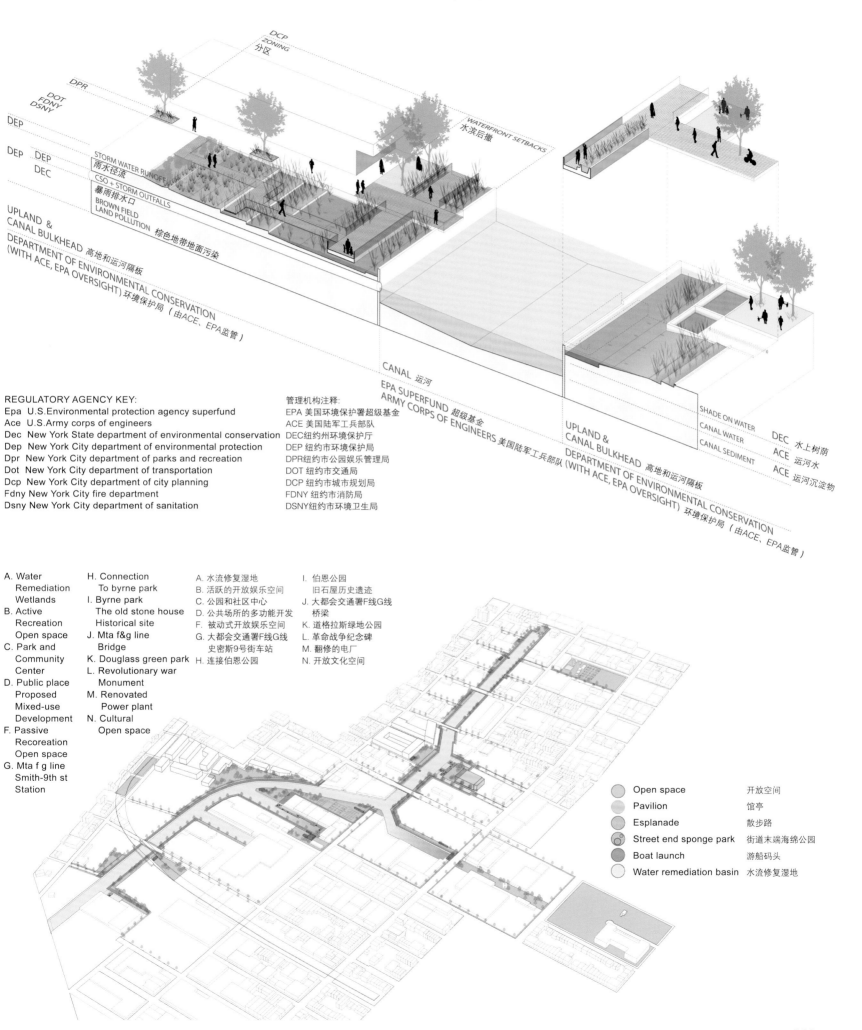

"Low Impact Development: A Design Manual for Urban Areas"
《低影响开发：城区规划手册》

Jury Comments:
What a useful, easy-to-understand tool, not only for architects, but also for community leaders and citizens working to ensure environmentally sustainable development. A very clear manual that should become the primer for creating beautiful and sustainable public streets and spaces. Urban design at a scale that architects can grasp and incorporate into their own projects. It is a project both specifically technical and inspiring all at the same time.

评委评语：
这是一个实用而通俗易懂的工具，不仅针对建筑师，还对社区领导人和有意向了解环保可持续开发的市民大有裨益。
这是一本清晰的手册，应该成为打造美观的可持续公共街道和空间的引导读本。
手册介绍了建筑师能够掌握并运用到自己项目中的城市规划，既独特又具有启发性。

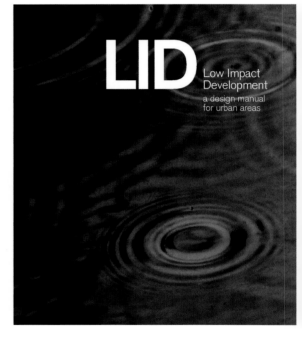

Notes of Interest

"Low Impact Development: a design manual for urban areas" is a 230-page publication designed for use by those involved in urban development-from homeowners, to institutions, developers, designers, cities, and regional authorities. Low Impact Development (LID) is an ecologically-based stormwater management approach favoring soft engineering to manage rainfall on site through a vegetated treatment network. The objective is to sustain a site's pre-development hydrological regime by using techniques that infiltrate, filter, store, and evaporate stormwater runoff close to its source.

The manual presents a graphic argument, using urban design templates and scenarios to illustrate the role of LID technology in regional planning and infrastructural design. It is the first to devise a LID Facilities Menu of the 21 assemblages available, organized from mechanical to biological functioning, and based on increasing level of treatment service (quality) and level of volume reduction service (quantity).

The manual's unique contribution to the topic lies in its advancement of LID from a set of Best Management Practices to a highly distributed treatment network deployed at neighborhood, municipal, and regional scales. It shifts LID from an isolated technology to a planning model based on nestled pattern languages in making places. The goals are to promote implementation of LID technologies in urban areas through adoption of best practices in planning and design, and encourage reform in municipal codes.

Engineer: University of Arkansas Ecological Engineering Group
Funding: The Arkansas Natural Resources Commission (ANRC) through a grant from the United States Environmental Protection Agency (USEPA) Region 6 Section 319(h)
Publication: Made possible by generous support from:
Arkansas Forestry Commission's Urban,
Forestry Program and US Forest Service Beaver Water Distri
Community Foundation of the Ozarks and Stewardship Ozark Initiative,
Ozarks Water Watch with Upper White River Basin Foundatio
National Center for Appropriate Technology Southeast Field Office: Fayetteville, Arkansas,
US Green Building Council Western Branch Arkansas Chapte
Northwest Arkansas,
Illinois River Watershed Partnership

工程师： 阿肯色大学生态工程集团
资金来源： 美国环境保护署6区319部门为阿肯色自然资源委员会提供的拨款
出版： 阿肯色林业委员会的城市林业项目和美国林业局海狸水地区分局、欧扎克社区基金会和欧扎克管理方案、
欧扎克水表和上白河流域基金会、
美国适用技术中心东南区办公室：阿肯色州，费耶特维尔、
美国绿色建筑委员会西部分会阿肯色分会：阿肯色州西北部、
伊利诺伊河流域集团

Architect / 建筑师	**Location** / 项目地点	**Photo Credit** / 图片版权
University of Arkansas Community Design Center 阿肯色大学社区设计中心	Fayetteville, Arkansas 阿肯色州,法耶特维尔	© University of Arkansas Community Design Center 阿肯色大学社区设计中心

LID IS SCALABLE TO BUILDING, PROPERTY, STREET, AND OPEN SPACE SYSTEMS.
低影响开发适用于建筑、地产、街道以及开放空间系统

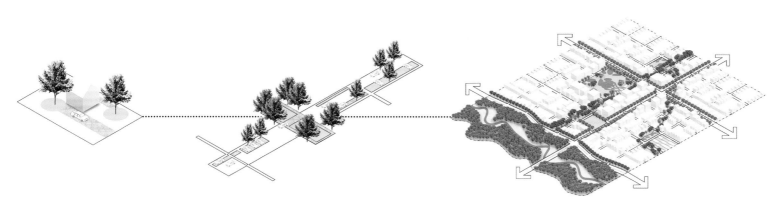

Lots: LID lots infiltrate stormwater through reduction or elimination of impervious surfaces and replacement of turf grass with productive landscapes.
土地:低影响开发土地减少或除去非渗透性地面并以多产的草皮替代,从而达到令雨水渗入的效果。

Streets: LID streets are green streets reducing and filtering runoff as it enters public space while enhancing the quality of place.
街道:低影响开发街道是能减少或过滤雨水径流的绿色街道,它们能提升空间的质量。

Networks: LID networks contain treatment facilities connected to regionally scaled systems of stormwater management.
网络:低影响开发网络包含着与区域雨水系统相连的处理设施。

项目特色

《低影响开发:城区规划手册》是一本230页的出版物,专门为参与城市开发的相关人士(例如:私房屋主、机构、开发商、设计师、城市和区域管理部门)设计。低影响开发是一种生态环保的雨水处理方式,倾向于通过植被处理网络柔和地处理场地上的雨水。其目标是通过渗透、过滤、存储和蒸发雨水径流来保持场地的开发前水文状况。

《手册》呈现了图表论证,利用城市规划模板和情境来展示了低影响开发在区域规划和基础设施设计中的重要性。它首次通过21个实例展示了低影响开发设施选单。项目从机械到生态功能进行排列,以增强处理服务(质量)和减容服务(数量)的层次为基础。

《手册》通过一系列最佳处理实例展示了先进的低影响开发技术,涉及范围包括社区、市政乃至大型区域。它将低影响开发从独立的技术转化为区域规划模型。它的目标是通过规划设计实例提升低影响开发技术在城区规划中的实施性并且独立市政法规的改革。

4 地下滞留 / underground detention

1. Parking
2. Inlet drain
3. Detention cell
4. Outlet pipe

1. 停车场
2. 入口
3. 滞留格
4. 排水管

Optimal level of service:
Detention/infiltration
最佳服务层次：
滞留/渗透

Location in LID network:
Optimally placed after filtration facilities
To prevent excessive sedimentation
在无影响开发网络中的位置：
在渗透设施之后，防止过多的沉淀

Scale:
Maximum watershed runoff area is 25 acres
规模：
最大流域为25英亩（约101,171平方米）

Management regime:
Inspection and sediment cleanout
管理体制：
检查和沉淀清洁

REFERENCES:
Low Impact Development Manual for Michigan
Urban Design Tools – Low Impact Development
Minnesota Urban Small Sites BMP Manual
参考文献：
《密歇根低影响开发手册》
《城市设计工具——低影响开发》
《明尼苏达州小型城市场地开发手册》

UNDERGROUND DETENTION

Underground detention systems detain stormwater runoff prior to its entrance into a conveyance system.
Underground storage systems store and slowly release runoff into the LID network. Some Systems can infiltrate stormwater if the soil beneath is permeable. Underground storage is employed in places where available surface area for ongrade storage is limited.
Underground storage reduces peak flow rate through metered discharge and has potential for infiltration. Improved water quality is achieved by sedimentation, or the setting of suspended solids. Though at first costly, underground detention systems are easy to access and maintain.

地下滞留

地下滞留系统在雨水径流进入传输系统之前对其进行扣留。
地下存储系统存储并缓慢释放雨水径流进入无影响开发网络。如果土壤具有渗透性，一些系统能够渗透雨水。地下存储系统用于地面存储不足的区域。
地下存储系统通过可测量的排放量来减少洪峰流量，并具有渗透潜力。沉淀或设置悬浮固体物能够提升水的质量。尽管初期成本较高，地下滞留系统便于接入和维护。

LID NETWORKS OFFER THE FULL RANGE OF ECOSYSTEM SERVICES
无影响开发网络提供全方位的生态系统服务

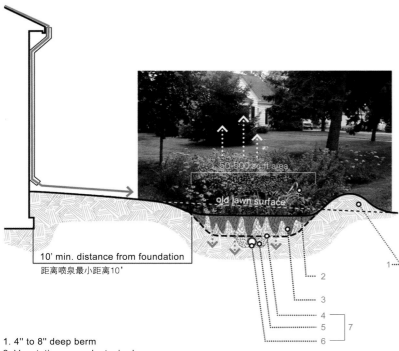

1. 4" to 8" deep berm
2. Vegetation: succulents, herbs, grasses
3. Amended soil mix
4. Filter fabric
5. 3/4" gravel base
6. Perforated underdrain
7. Overflow system for poorly-drained soils or large storm events

1. 4"–8"深的路肩
2. 植物：多浆植物、草本植物、草
3. 改进的土壤配置
4. 过滤构造
5. 3/4"碎石底层
6. 多孔排水管
7. 为排水性差的土壤或大型暴雨所设计的上溢系统

10' min. distance from foundation
距离喷泉最小距离10'

REFERENCES:
Low Impact Development Design Strategies – An Integrated Design Approach
Low Impact Development Manual for Michigan
Low Impact Development Technical Guidance Manual for Puget Sound
United States Department of Housing and Urban Development
Minnesota Urban Small Sites BMP Manual
参考文献：
《低影响开发设计策略——综合设计方案》
《密歇根低影响开发手册》
《普吉特湾低影响开发技术指导》
《美国住宅和城市开发部》
《明尼苏达州小型城市场地开发手册》

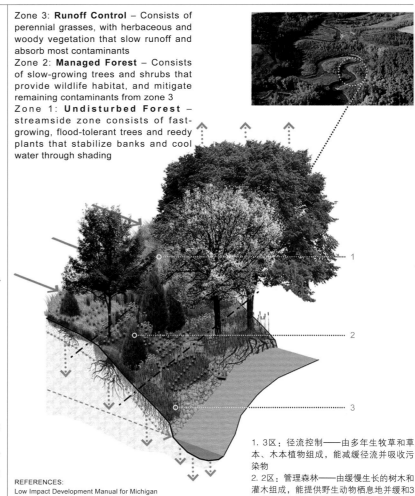

Zone 3: **Runoff Control** – Consists of perennial grasses, with herbaceous and woody vegetation that slow runoff and absorb most contaminants
Zone 2: **Managed Forest** – Consists of slow-growing trees and shrubs that provide wildlife habitat, and mitigate remaining contaminants from zone 3
Zone 1: **Undisturbed Forest** – streamside zone consists of fast-growing, flood-tolerant trees and reedy plants that stabilize banks and cool water through shading

1. 3区：径流控制——由多年生牧草和草本、木本植物组成，能减缓径流并吸收污染物
2. 2区：管理森林——由缓慢生长的树木和灌木组成，能提供野生动物栖息地并缓和3区残留下来的污染物
3. 1区：原始森林——河滨地带，由快速生长的抗洪树木和芦苇植物组成，能稳固河岸并通过树荫冷却水

REFERENCES:
Low Impact Development Manual for Michigan
Conservation Buffers: Design Guidelines for Buffers, Corridors, and Greenways
参考文献：
《密歇根低影响开发手册》
《缓冲区保护：缓冲区、走廊和林荫路的设计指导》

19 生态沼泽

bioswale

Optimal level of service:
Filtration/infiltration/treatment
最佳服务层次：
过滤/渗透/处理

Location in LID network:
Downstream of filtration components, But upstream of larger detention, Retention, or treatment facilities
在无影响开发网络中的位置：
过滤组件下游，大型滞留、扣留或处理设施的上游

Scale:
2'-8' wide with 2"-4" optimal water depth
规模：
2'-8'宽，2''-4''最佳水深

Management regime:
Occasional removal of trash and pruning of vegetation
管理体制：
偶尔处理垃圾和修剪植物

REFERENCES:
Low Impact Development Design Strategies – An Integrated Design Approach
Low Impact Development Manual for Michigan
Low Impact Development Technical Guidance Manual for Puget Sound
Unite States Department of Housing and Urban Development
Minnesota Urban Small Sites BMP Manual
参考文献：
《低影响开发设计策略——综合设计方案》
《密歇根低影响开发手册》
《普吉特湾低影响开发技术指导手册》
《美国住宅和城市开发部》
《明尼苏达州小型城市场地开发手册》

1. Vegetation: succulents, herbs, grasses
2. Mulch layer, 2"-3" with 1: 3 slope or less
3. Overflow grate
4. Amended soil mix: typically 45% sand, 35% top soil, and 20% compost
5. Filter fabric
6. 3/4" gravel base
7. Perforated underdrain
8. overflow system for poorly-drained soils or large storm events

1. 植物：多浆植物、草本植物、草
2. 覆盖层，2"-3"，带有1: 3的斜坡或更小坡度的斜坡
3. 上溢格栅
4. 改进的土壤配置：45%沙，35%表层土，20%堆肥
5. 过滤构造
6. 3/4''碎石底层
7. 多孔排水管
8. 为排水性差的土壤或大型暴雨所设计的上溢系统

BIOSWALE

A bioswale is an open, gently sloped, vegetated channel designed for treatment and conveyance of stormwater runoff. Bioswales are a bioretention device in which pollutant mitigation occurs through Phytoremediation by facultative vegetation. Bioswales combine treatment and Conveyance services, reducing land development costs by eliminating the need. For costly conventional conveyance systems. The main function of a rain garden is to treat stormwater runoff as it is infiltrated. Bioswales are usually located along roads, drives, or parking lots where the contributing acreage is less than five acres.
Bioswales require curb cuts, gutters or other devices that direct flow to them. They may require an underdrain where soil permeability is limited, as well as an overflow grate for larger storm events.

生态沼泽

生态沼泽是一个开放的缓坡式植被管道，专门为处理并传输雨水径流而设计。
兼生性植物在生态沼泽中进行植物修复，分解污染物。生态沼泽结合了处理和输送功能，通过消除昂贵的传统传输系统来减少土地开发成本。雨水花园的主要功能室处理雨水径流，使其渗透到地下。生态沼泽通常沿着公路、车道或者停车场而建，总面积在5英亩（约20,234平方米）以下。
生态沼泽需要斜坡、水沟或其他设备来引导水流。它们需要一个限制土壤渗透的地下排水管和一个上溢排架来应对大型暴雨。

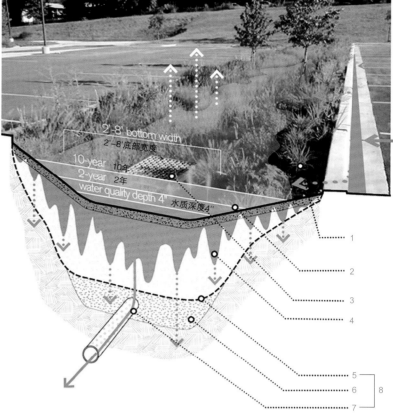

21 人工湿地

constructed wetland

Optimal level of service:
Retention/filtration/infiltration/treatment
最佳服务层次：
扣留/过滤/渗透/处理

Location in LID network:
End-of-line facility, upstream of overflow basins or receiving water bodies
在无影响开发网络中的位置：
终端设施，溢流池或接受水体的上游

Scale:
From pocket wetlands managing up to 10 acres of drainage to shallow marshes managing more than 25 acres of drainage
规模：
从10英亩（约4.05公顷）的迷你湿地排水系统到25英亩（约10.1公顷）的浅层沼泽排水系统

Management regime:
System requires removal of trash and sediment between two and ten years, and semiannually during first three years
管理体制：
系统需要在2到10年之间处理垃圾和沉淀物，前3年每半年需要处理一次。

REFERENCES:
Low Impact Development Manual for Michigan
United States Department of Housing and Urban Development
Minnesota Urban Small BMP Manual
参考文献：
《密歇根低影响开发手册》
《美国住宅和城市开发部》
《明尼苏达州小型城市场地开发手册》

1. Treatment zone
2. Sedimentation zone
3. Facultative vegetation
4. 12" native topsoil
5. Filter fabric
6. Water table

1. 处理区
2. 沉淀区
3. 兼生性植被
4. 12''原生表层土
5. 过滤设备
6. 潜水面

CONSTRUCTED WETLAND

Constructed wetlands are artificial marshes or swamps with permanent standing water that offer a full range of ecosystem services to treat polluted stormwater.
Considered to be a comprehensive treatment system, constructed wetlands, like infiltration basins, require intrinsic hydrogeologic properties to reproduce natural watershed functioning. As with other infiltration system, resulting in eutrophication or an oxygen deprived system.
Constructed wetlands are land rich biofilters and differ from retention ponds in their shallower depths, greater vegetation coverage, and extensive wildlife habitat. They require relatively large contributing drainage areas to maintain a shallow permanent pool. Minimum contributing drainage area should be at least 10 acres, although pocket wetlands may be appropriate for smaller sites if sufficient water flow is available.

人工湿地

人工湿地是人造沼泽或湿地，具有永久性的静水，能利用全方位的系统服务来处理被污染的雨水。
作为一个综合处理系统，人工湿地与渗透池一样，需要固有的水文地质性能来再造天然流域功能。与其他渗透系统一样，会导致富营养化或缺氧系统。
人工湿地是地面丰富的生态滤池，与澄清池不同，它们拥有更浅的深度、更大的植被覆盖率和更多的野生动植物栖息地。它们需要相对大的排水面积来保持浅层永久水池。最小的排水区域至少应该为10英亩，尽管迷你湿地可能更适合水流充足的小型场地。

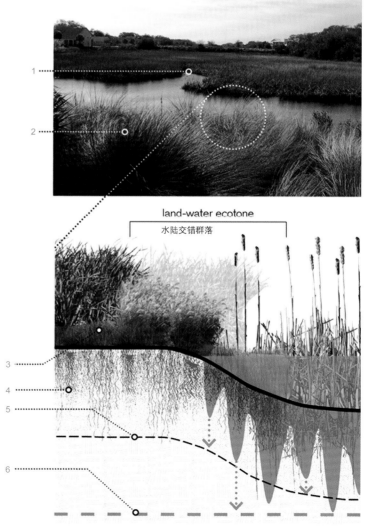

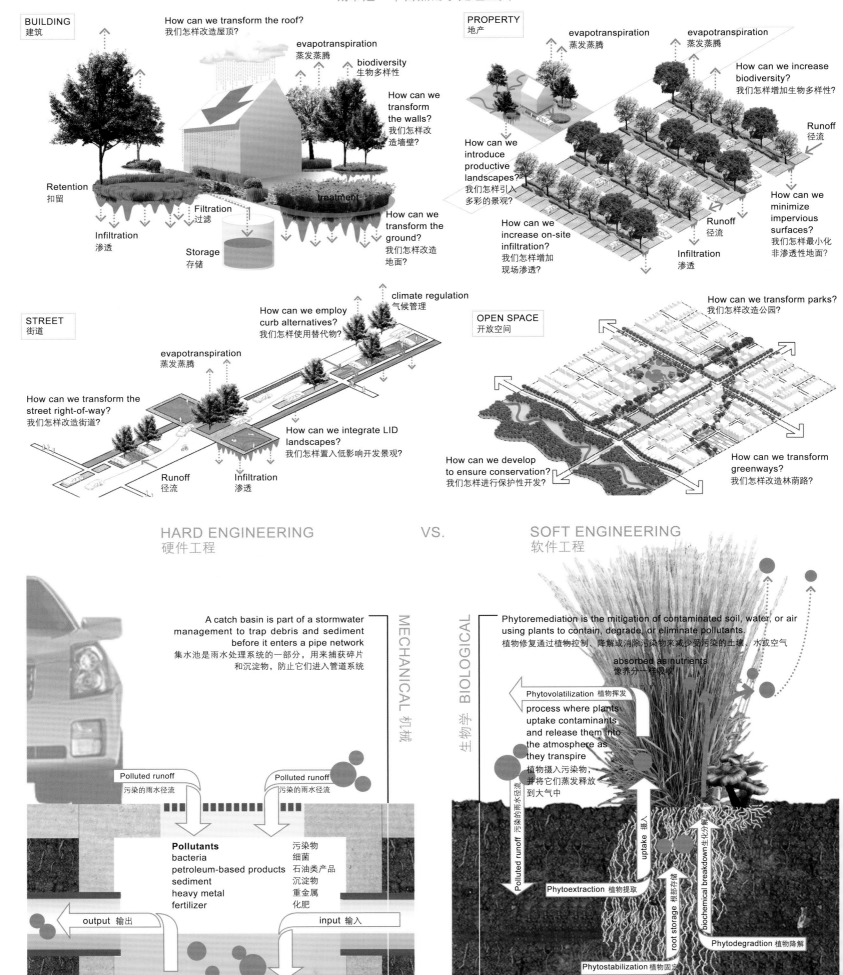

THE SIX WATER TREATMENT TECHNOLOGIES
六种水处理技术

Mechanical 机械 — Biological 生物学

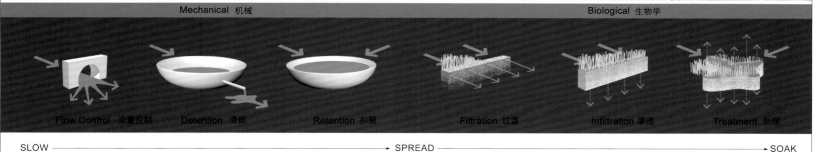

Flow Control 流量控制 | Detention 滞留 | Retention 扣留 | Filtration 过滤 | Infiltration 渗透 | Treatment 处理

SLOW 缓慢 ——————— SPREAD 扩展 ——————— SOAK 浸透

flow control: The regulation of Stormwater runoff flow rates.
流量控制：控制雨水径流的流速。

detention: The temporary storage of stormwater runoff in underground areas to allow for metered discharge that reduce peak flow rates.
滞留：在地下短暂存储雨水径流，以测量减少最大流速的排放量。

retention: The storage of stormwater runoff on site to allow for sedimentation of suspended solids.
扣留：当场存储雨水径流，沉淀固体悬浮物。

filtration: The sequestration of sediment from stormwater runoff through a porous media such as sand, a fibrous root system, or a man-made filter.
过滤：通过多孔的媒介（例如沙子、须根系统或人工过滤器）将沉淀物与雨水径流隔离。

infiltration: The vertical movement of stormwater runoff through soil, recharging groundwater.
渗透：雨水径流穿透土壤的垂直运动，重新补充地下水。

treatment: Processes that utilize phytoremediation or bacterial colonies to metabolize contaminants in stormwater Runoff.
处理：利用植物修复或细菌群落来代谢雨水径流里的污染物。

THE 21 LID FACILITIES
21个低影响开发设施

1. Oversized pipes
2. Flow control devices
3. Dry swale
4. Underground detention
5. Detention pond
6. Wet vault
7. Rainwater harvesting
8. Retention pond
9. Filter strip
10. Underground sand filter
11. Surface sand filter
12. Vegetated wall
13. Vegetated roof
14. Pervious paving
15. Infiltration trench
16. Tree box filter
17. Rain garden
18. Riparian buffer
19. Bioswale
20. Infiltration basin
21. Constructed wetland

1. 特大型管道
2. 流量控制设备
3. 干沼泽
4. 地下滞留
5. 滞留池
6. 水库
7. 雨水收集
8. 扣留池
9. 过滤带
10. 地下沙过滤
11. 表面沙过滤
12. 植物墙
13. 植物屋顶
14. 可渗透铺装
15. 下渗沟
16. 树下箱过滤
17. 雨水花园
18. 滨水缓冲带
19. 生态沼泽
20. 渗透盆地
21. 人造湿地

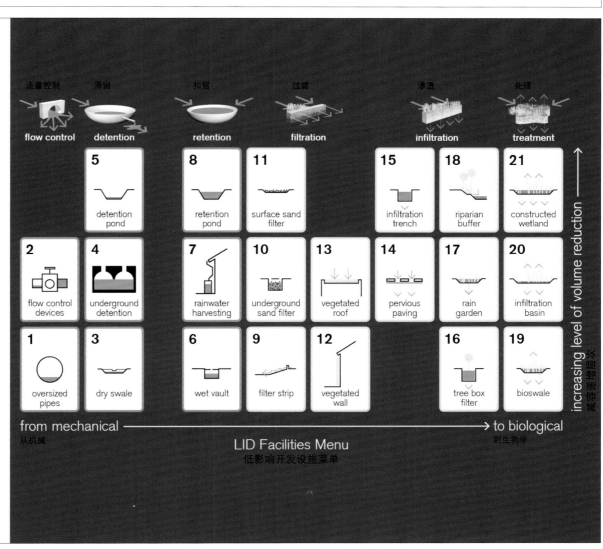

LID Facilities Menu
低影响开发设施菜单

Townscaping an Automobile-Oriented Fabric
以机动车为主的城市网格中的城市景观规划

Jury Comments:
There is much that we can learn from smaller communities, and "townscaping" is a creative example of what a small, long-established community can do to transform its 20th century roadway system into a 21st-century amenity. An urban design approach that is both design driven and community oriented simultaneously. This plan proves that a place laid out originally for cars can be adapted to a future where people are connected in other ways. A beautiful model for greening and organizing small town USA.

评委评语：
我们可以从小型住宅区中获益良多，"城市景观规划"是小型长期住宅区的创意设计典范，成功地将20世纪公路系统转化为21世纪基础设施。
这一城市规划兼顾了设计与地区要求，证明了一个汽车区域是能够被改造成适宜行人行走的步行区的。
美国绿化设计和小城规划的典范。

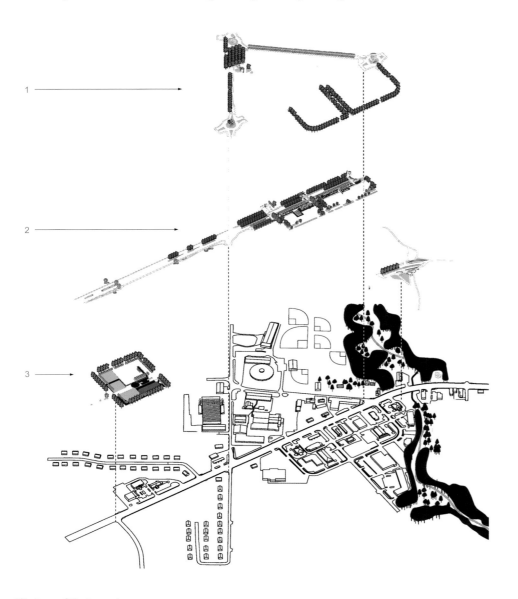

1. STEP3: Accentuate with landmarks
2. STEP2: Frame with landscape and new street geometries
3. STEP1: Bound with gateways
1. 第三步：通过地标来突出地点
2. 第二步：通过景观和新的街道几何图形来建造框架
3. 第一步：通过大门来界定区域

Notes of Interest
The townscape plan for Farmington proposes new public landscapes to restitch a 5,000-person bedroom community fragmented by a five-lane commercial arterial. Once a vibrant farming community, central to one of the nation's largest strawberry and apple-producing regions in the early 1900s, Farmington is now a bedroom community. Unlike the totalizing pattern of a master plan, townscaping employs a serial organization of nodes to create a walkable urban environment within an automobile-oriented fabric.
As a retrofit planning strategy, townscaping offers a model for an incremental urbanization without reliance on capital-intensive architectural investments. The goal is to create a memorable town fabric for anchoring new growth in an otherwise

Funding: The National Endowment for the Arts 2010 Grant Award for Access to Artistic Excellence
Owner: City of Farmington
资金来源：国家艺术基金会
所有人：法明顿市

Architect / 建筑师
University of Arkansas Community Design Center
阿肯色大学社区设计中心

Location / 项目地点
Farmington, Arkansas
阿肯色州，法明顿

Photo Credit / 图片版权
© University of Arkansas Community Design Center
阿肯色大学社区设计中心

1. Farmers market
2. Town green
3. Hard fruit trees
4. Espalier
5. Hard fruit trees
6. Soft fruit trees
7. Soft fruit orchard
8. Air filtration
9. Hard fruit allee
10. Soft fruit allee
11. Arboretum
12. Pedestrian
13. Arcade
14. Roosting Tower
15. Foraging riparian
16. Viticetum Gateway

1. 农贸市场
2. 城镇绿地
3. 硬果树
4. 树墙
5. 硬果树
6. 软果树
7. 软果园
8. 空气过滤
9. 硬果大道
10. 软果大道
11. 树木园
12. 人行道
13. 拱廊
14. 栖息塔
15. 觅食河岸
16. 藤本植物大门

fragmented and anonymous landscape. Working within the context of Farmington's limited resources, townscaping creates "articulated environments" through sleuth planning techniques.

Beginning with the ordinary components commonly budgeted in urban infrastructure, the townscape plan condenses these elements into a series of nodes that galvanize a sense of place. The townscape plan for Farmington integrates multiple placemaking strategies in: 1) context-sensitive highway design, 2) public art planning, and 3) agricultural urbanism. Placemaking in the townscape vocabulary offers a strategic pedestrianization of automobile-oriented patterns without denying the automobile's fundamental role in servicing contemporary development.

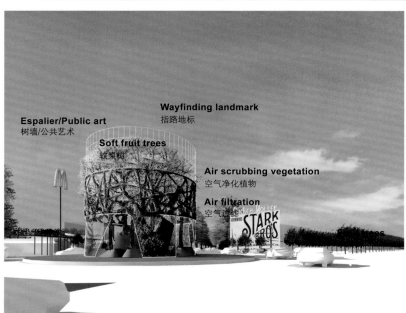
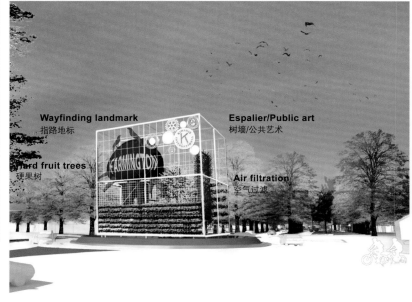

NEW MULTI-WAY BOULEVARD AND PEDESTRIAN INTERFACE
新多向大道和步行街接口

The new multi-way boulevard can accommodate existing land uses while incentiving new urban mixed-use development.
新多向大道能够调解现有的土地使用同时刺激新城市综合项目的开发。

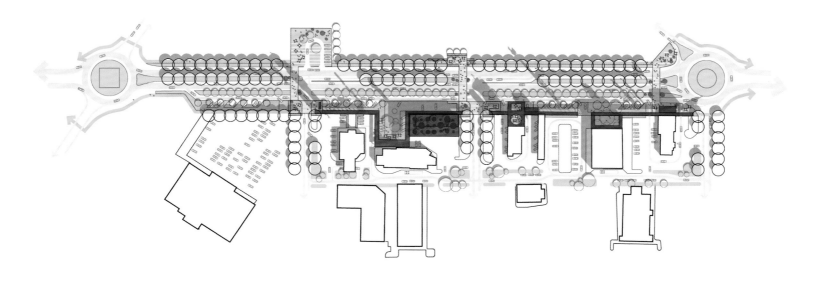

项目特色

法明顿城市景观规划旨在为一个由五车道商业主干道环绕的5,000人近郊住宅区打造全新的公共景观。这个住宅区在20世纪早期曾是一个生机勃勃的农业区域，地处美国最大的草莓和苹果产地中央。与整体城市规划不同，城市景观绘画采用了一系列组织节点在以机动车为主的城市网格中打造适用步行的城市环境。

作为一个可行性规划策略，城市景观规划提供了不依赖密集资金建筑投资的增值城市模型。其目标是打造一个令人难忘的城市网络并建立全新的连续景观设施。在法明顿有限的资源条件下，城市景观规划通过检查规划技术打造了"连贯的环境"。

规划采用了城市设施中的普通元素，并将其浓缩到一系列空间节点之中。城市景观规划结合了多重空间策略：1)环境敏感高速公路设计；2)公共艺术规划；3)农业化城市主义。规划中的区域设计在以机动车为主的环境中打造了一个策略性步行区，同时又不会否认机动车在现代城市开发中的基础作用。

John Hancock Tower
约翰·汉考克大厦

Notes of Interest

The John Hancock Tower is an office building commissioned by the John Hancock Mutual Life Insurance Company principally for its own use, with a few floors occupied by other tenants. The building contains a gross floor area of 2,060,000 square feet on sixty-two floors above grade and two below. The architect was I. M. Pei & Partners (now Pei Cobb Freed & Partners), with Henry N. Cobb, FAIA, as design partner. Construction was completed in 1976.
The site's adjacent architectural landmarks – in particular Trinity Church and the Boston Public Library – were what made Cobb's mission exceptionally difficult. It became apparent that it would be a great challenge to adhere to the firm's trademark rationalism and insert such a tall building into the site without fatally rupturing Copley Square's sense of scale and proportion.
The solution was single-minded and rested in the minimalism of the design: a smooth, reflective, glass tower with no spandrel panels and minimal mullions – essentially a very large mirror. To minimize its intrusion on the adjacent landscape, the building is rhomboid in shape and placed diagonally on the site, so its shorter, slightest side faces the church and plaza.
The John Hancock Tower recently achieved LEED Gold Existing Building certification for energy use, lighting, water, and material use as well as a variety of other sustainable strategies. Some of these involve equipment upgrades, while others were integral to the original design. For example, the building's glass façade and narrow floor plate allow natural light to reach 86 percent of all work areas.
In awarding it the 1983 Harleston Parker Medal, the Boston Society of Architects jury unanimously agreed that the John Hancock Tower met its criterion: to be the city's "most beautiful piece of architecture". Paul Goldberger, Hon. AIA, architecture critic for the The New Yorker, went even further when he wrote recently, "the John Hancock Tower remains one of the most beautiful skyscrapers ever built."

项目特色

约翰·汉考克大厦由约翰·汉考克人寿保险公司委托建造，主要用于自用办公，有部分楼层出租。大厦总楼面面积191,380平方米，地上62层，地下2层。大厦完成于1976年，由贝聿铭建筑事务所和亨利·N·科博（美国建筑师协会会员）共同设计。
建筑场地紧邻地标性建筑——三一教堂和波士顿公共图书馆，这让建筑师科博的工作变得异常艰难。坚持公司商标的理性主义并在柯普利广场区域插入一个如此高的大厦，同时又不能破坏广场的整体比例和规模——这无疑是一项严峻的挑战。
该项目的解决方案十分专一，依赖于设计的极简主义特色：光滑而具有反射性的玻璃大厦没有采用任何拱肩结构，并且将窗框设计最小化——看起来就是一面巨大的镜子。为了减轻它对周边景观的破坏，大厦采用了偏菱形造型，以对角线形式设置在场地上，让自身较短并且不重要的一面朝向教堂和广场。
由于在能源、照明、水和材料等方面对可持续策略的运用，约翰·汉考克大厦最近获得了绿色建筑金奖认证。一些相关设备升级，另一些则被纳入最初的计划之中。例如，建筑的玻璃外立面和狭窄的楼面让自然光覆盖了86%的办公区域。在1983年大厦获得哈尔勒斯顿·派克奖之际，波士顿建筑协会的评委们一致同意约翰·汉考克大厦满足了奖项标准：认为它是城内"最美丽的建筑"。保罗·哥德伯格（美国建筑师协会荣誉会员；《纽约客》特约建筑评论员）最近写到："约翰·汉考克大厦是最美丽的摩天大楼之一。"

Architect /建筑师
I.M. Pei & Partners (now Pei Cobb Freed & Partners)
贝聿铭建筑事务所（现名：PCF建筑事务所）

Location /项目地点
Boston
波士顿

Photo Credit /图片版权
© Gorchev & Gorchev; Peter Vanderwarker (Detail)
格尔奇夫&格尔奇夫、彼得·范德瓦尔克（细部）

2012 Institute Honor Awards

2012年美国建筑师协会建筑/室内设计/区域和城市规划荣誉奖评委

2012 INSTITUTE HONOR AWARDS FOR ARCHITECTURE JURY
建筑荣誉奖评委

Rod Kruse, FAIA, Chair
BNIM Architects
洛德·克鲁斯
美国建筑师协会会员；评委会主席
BNIM 建筑事务所

Barbara White Bryson, FAIA
Rice University
芭芭拉·怀特·布莱森
美国建筑师协会会员
莱斯大学

Annie Chu, AIA
Chu & Gooding Architects
安妮·朱
美国建筑师协会
朱&古丁建筑事务所

Dima Daimi, Assoc. AIA
Rossetti
迪玛·达依米
美国建筑师协会
罗塞蒂公司

Harry J. Hunderman, FAIA
Wiss, Janney, Elstner Associates, Inc.
哈里·J·哈德曼
美国建筑师协会会员
WJS 建筑事务所

Scott Lindenau, FAIA
Studio B Architects
斯科特·林德诺
美国建筑师协会会员
B 建筑工作室

Kirsten R. Murray, AIA
Olson Kundig Architects
科斯顿·R·马雷
美国建筑师协会
奥尔森·昆丁建筑事务所

Thomas M. Phifer, FAIA
Thomas Phifer & Partners
托马斯·M·费佛
美国建筑师协会会员
托马斯·费佛建筑事务所

Seth H. Wentz, AIA
LSC Design, Inc.
塞斯·H·文兹
美国建筑师协会
LSC 设计公司

2012 INSTITUTE HONOR AWARDS FOR INTERIOR ARCHITECTURE JURY
室内设计荣誉奖评委

Elizabeth Corbin Murphy, FAIA, Chair
CMB Architects
伊利莎白·科尔宾·墨菲
美国建筑师协会会员；评委会主席
CMB 建筑事务所

Robert Allen, Jr., AIA
Metalhouse
罗伯特·艾伦
美国建筑师协会
金属房子公司

Mark Jensen, AIA
Jensen Architects
马克·詹森
美国建筑师协会
詹森建筑事务所

David Lenox, AIA
University Architect/Dir. Campus Planning
Stanford University
大卫·雷诺克斯
美国建筑师协会
大学建筑师 / 校园规划方向
斯坦福大学

Erick S. Ragni, AIA
MaRS Architects
埃里克·S·拉格尼
美国建筑师协会
马尔斯建筑事务所

2012 INSTITUTE HONOR AWARDS FOR REGIONAL AND URBAN DESIGN JURY
区域和城市规划荣誉奖评委

Bruce Lindsey, AIA, Chair
Washington University in St. Louis
布鲁斯·林赛
美国建筑师协会；评委会主席
华盛顿大学圣路易斯分校

Catherine Seavitt Nordenson, AIA
Catherine Seavitt Studio
凯瑟琳·西维特·诺德森
美国建筑师协会
凯瑟琳·西维特工作室

Martha Welborne, FAIA
Los Angeles County Metropolitan Transportation Authority
马尔撒·威尔伯恩
美国建筑师协会会员
洛杉矶县都市运输局

Rod Kruse, FAIA, LEED AP
2012 Chair,
Institute Honor Awards for Architecture
洛德·克鲁斯
美国建筑师协会会员；美国绿色建筑协会认证专家
2012 年美国建筑师协会建筑荣誉奖评委会主席

Elizabeth Corbin Murphy, FAIA
2012 Chair,
Institute Honor Awards for Interior Architecture
伊利莎白·科尔宾·墨菲
美国建筑师协会会员
2012 年美国建筑师协会室内设计荣誉奖评委会主席

© Mary S. Watkins 玛丽·S·沃特金斯

© Studio Martone 马尔通工作室

Rod Kruse, Principal | BNIM has built a reputation as one of the Central States Region's strongest design talents. He has gained recognition on national, regional and local levels for his projects, including a 2002 National AIA Award for Design Excellence for Architecture for the Newton Road Parking and Chilled Water Facility at the University of Iowa and a 2000 National AIA Award for Design Excellence for Architecture for the Center Street Park and Ride Facility. In the role of Principal, he is uniquely honored by having received two AIA National Firm Awards, for BNIM in 2011 and for Herbert Lewis Kruse Blunck Architecture in 2001.

Well-established as a leader in design, Rod's work has been included in several traveling exhibitions, and he has lectured widely. Rod's work and writings have also been featured in numerous national and regional periodicals including Architecture, Architectural Record, I.D. Magazine and Iowa Architect. Rod has chaired or served on numerous Design Excellence Awards juries and also chaired the AIA College of Fellows jury.

Recognizing Rod's Design Excellence and Leadership, the American Institute of Architects named him a Fellow 1996.

洛德·克鲁斯（BNIM 设计公司总监）被誉为是美国中部地区最好的设计师。他的项目获得了全国、全地区以及本地层面的认证。其中，纽顿路停车场和爱荷华大学冷水处理设施获得了 2002 美国建筑师协会优秀建筑奖，中央街道公园和换乘设施获得了 2000 年美国建筑师协会优秀建筑奖。作为事务所总监，他获得了两个美国建筑师协会事务所奖，一次是 2011 年的 BNIM 事务所，一次是 2001 年的 HLKB 建筑事务所。

作为一名出色的设计总监，洛德的作品已经在几次巡回展览中进行了展示，他也到四处进行了讲座。洛德的作品和文章在大量的国家和区域性期刊中出现，其中包括《建筑》《建筑实录》《I.D. 杂志》和《爱荷华建筑师》。洛德多次主持并参与了优秀设计奖的评选工作，同时也是美国建筑师协会学院研究员的评委会主席。

由于洛德的优秀设计和领导才能，美国建筑师协会在 1996 年授予他资深会员资格。

Chambers, Murphy & Burge will recycle old historic structures for new uses or restore unique landmarks to strict conservation standards. The firm of which Ms. Murphy is principal was founded as Chambers & Chambers Architects almost fifty years ago, and is dedicated solely to preservation and restoration technology and design. Elizabeth and her business partner consult with other architects on projects like the Dallas County Courthouse, the Supreme Court Building in Columbus, and the Minnesota State Capitol. The firm assists architects and building owners with the state and federal rehabilitation tax credits, design related to old or historic structures, detailed restoration specifications, and historic interiors. Elizabeth has developed with Edsel & Eleanor Ford House a Cyclical Maintenance Plan for the care of the six Albert Kahn structures at Gaukler Point. The Cyclical Maintenance Plan (interior and exterior materials) for the Henry Ford Estate in Dearborn incorporates 13 structures, in a Jens Jensen landscape.

Elizabeth Corbin Murphy, FAIA is past chair of the Advisory Group for the American Institute of Architects National Committee on Historic Resources. Ms. Murphy is a Professional Peer for the GSA Design Excellence and First Impressions Programs. She has served on several design awards juries including the AIA National Honor Awards, the GSA National Design Excellence Awards, the AIA Ohio Gold Medal Award, the AIA Cleveland Honor Awards, and the Charles E. Peterson Prize Awards. Having served for nine years with the Board of Regents for the American Architectural Foundation and serving on the Octagon Committee, Ms. Murphy now serves on the Executive Board for AIA Ohio and the Board of Advocates for the Preservation Institutes of Nantucket (PI:N) and Saint Augustine (PI:SA), both programs of the University of Florida College of Design Construction and Planning.

With her professional degree from the University of Notre Dame, Elizabeth completed the Masters of Architecture program at Kent State University to allow herself more research time in preservation of the built environment. Elizabeth is a Professor of Practice at Kent State University where the students she teaches have won eleven national awards in the fifteen entries made to the Charles E. Peterson Prize Competition. Ms. Murphy's students have more "first place" awards than any other professor. Ms. Murphy is registered with NCARB and with NCIDQ.

CMB 公司将重新利用旧建筑或者依据严格的保护标准对独特的地标性建筑进行修复。这个由伊利莎白·科尔宾·墨菲领衔的事务所原名为钱伯斯 & 钱伯斯建筑事务所，成立于 50 年前，一直致力于建筑修复保护技术与设计。伊利莎白和她的商业伙伴与其他建筑师一起探讨项目，例如达拉斯郡政府、哥伦比亚高级法院和明尼苏达州议会大楼。她的事务所帮助建筑师和建筑所有人处理州立和联邦政府的修复税额减免、进行旧建筑和历史建筑设计、建立细部修复规范以及进行历史建筑的室内设计。伊利莎白为艾德赛尔 & 埃莉诺·福特住宅开发了一套周期维护计划，用于保护"高克勒点"的 6 个艾伯特·卡恩结构。亨利·福特房产周期维护计划（包括室内和室外材料）包含延斯·延森景观中的 13 个结构。

伊利莎白是美国建筑师协会国家历史资源委员会的前任主席。墨菲女士是 GSA 优秀设计和第一印象项目的专业先锋。她担任多个设计奖的评委，其中包括美国建筑师协会国家荣誉奖、GSA 国家优秀设计奖、美国建筑师协会俄亥俄州金奖、美国建筑师协会克利夫兰荣誉奖和查尔斯·E·彼得森奖。墨菲女士曾经连续九年任职于美国建筑基金会评委会和八角委员会。现在，她任职于美国建筑师协会俄亥俄分会执行委员会和楠塔基特岛和圣奥古丁斯保护协会的拥护委员会。这两个项目都是佛罗里达大学建筑设计与规划学院运营。

在圣母大学获得了建筑学学士学位后，伊利莎白在肯特州立大学完成了建筑学硕士的学业，这样便有了更多的时间可以研究建成环境的保护工作。伊利莎白是肯特州立大学的实践教授，她所教授的学生在查尔斯·E·彼得森奖中提交了 15 件作品，并获得了 11 项国家大奖。她的学生所获的一等奖比其他教授的学生要多。伊利莎白获得美国建筑师注册委员会认证和美国室内设计师资格认证。

Bruce Lindsey, AIA
AIA 2012 Chair,
Institute Honor Awards for Regional & Urban Design
布鲁斯·林赛
美国建筑师协会
2012年美国建筑师协会区域和城市规划荣誉奖评委会主席

© Stan Strembecki 斯坦·斯特里姆贝奇

As Dean, E. Desmond Lee Professor for Community Collaboration, College of Architecture and Graduate School of Architecture & Urban Design, Washington University in St. Louis, Bruce Lindsey, AIA, has made significant contributions to beginning design education, sustainable design education, and community design education. He began his tenure as Dean of Architecture at Washington University in November 2006, and since then has led the Master of Landscape Architecture initiative, strengthened community design programs, and enhanced environmental education at all levels. Design Intelligence named him one of the Most Admired Educators of 2009.

Lindsey serves on the steering committees of the International Center for Advanced Renewable Energy and Sustainability and the Institute for Public Health, and the Gephardt Institute for Public Service. He also serves on the boards of the Downtown Partnership and the St. Louis chapter of the United States Green Building Council.

Lindsey's research has long focused on the application of digital tools to design and construction practice. In 1992, his work in digital-aided manufacturing was cited by Engineering News Record as one of the year's 10 most significant contributions to the construction industry. His book Digital Gehry: Material Resistance Digital Construction (2001) explores the use of technology in the design process of architect Frank Gehry. A practicing architect, Lindsey worked with Davis + Gannon Architects to design the Pittsburgh Glass Center, which earned a gold rating under LEED guidelines. The project also received an AIA Design Honor Award and was chosen as one of 2005's top 10 green buildings by the AIA's Committee on the Environment.

Lindsey earned an M. Arch degree at Yale University and an MFA in Sculpture & Photography and a BFA in Art from University of Utah.

作为华盛顿大学建筑学院和建筑设计与城市规划研究院院长，E·德斯蒙德·李教授称号的获得者。他在初期设计教育、可持续设计教育以及社区设计教育中做出了重大的贡献。他于2006年11月开始担任华盛顿大学建筑学院的院长，并自此发起了景观设计硕士学位计划，强化了社区设计项目，并因此在各个层面上促进了环境教育。"设计智慧"也将他列入了2009年最受尊敬的教育家名单。

林赛在国际可再生能源和可持续设计中心、公共卫生协会以及吉法尔特公共服务协会任职。他还是市区伙伴和美国绿色建筑协会圣路易斯分会的委员。

林赛的研究一直聚焦于数码工具在建筑和建设过程中的应用。1992年，他的数码辅助设计工作被《工程新闻纪录》评为年度十大建筑贡献。他所撰写的图书《数码盖里：抗材料数码建设》（2001）探索了弗兰克·盖里在设计过程中所运用的技术。作为一名执业建筑师，林赛与戴维斯+甘农建筑事务所共同设计了匹兹堡玻璃中心。该项目获得了绿色建筑金奖认证和美国建筑师协会设计荣誉奖，并且被美国建筑师协会的环境委员会评为2005年的十佳绿色建筑。

布鲁斯·林赛在耶鲁大学获得了建筑学硕士学位，并且在犹他州大学获得了雕塑与摄影艺术硕士学位以及艺术学士学位。

8 House
8字住宅

Jury Comments:
The 8 House masterfully recreates the horizontal social connectivity and interaction of the streets of a village neighborhood through a series of delightful accessible ramps in a mixed use, multifamily housing project. The skillful shaping of the mass of the facility provides an invigorating sculptural form while creating the ramped "pedestrian" street system and providing full depth dwelling units which are filled with light and views.
People really "live" in this newly created neighborhood with shopping, restaurants, an art gallery, office facilities, childcare, educational facilities and the sound of children playing. This is a complex and exemplary project of a new typology.

评委评语：
8字住宅巧妙地重塑了水平社会联系，通过一系列多功能坡道与乡村的街道形成了互动。建筑的造型具有雕塑感，创造了坡道步行街系统和具有良好视野和光线的深度住宅。
人们愉快地生活在这个新建成的街区中，享受着购物、餐厅、画廊、办公、儿童托管、教育和儿童游乐设施。这是一个全新的综合项目典范。

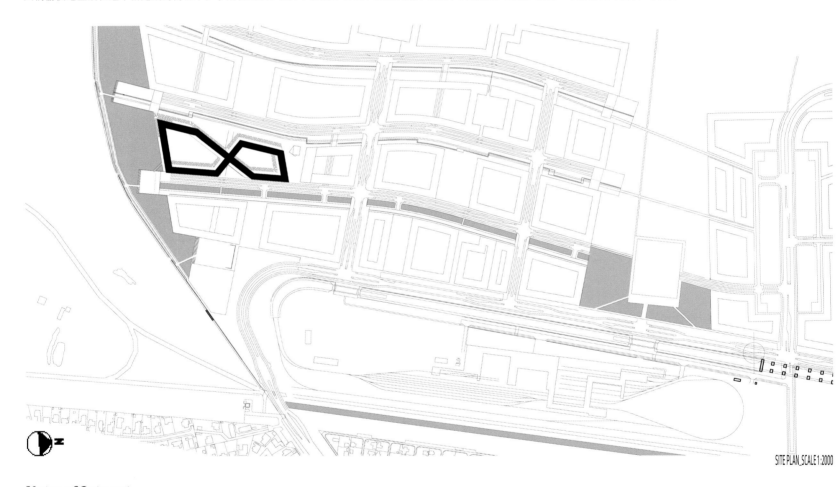

SITE PLAN_SCALE 1:2000

Notes of Interest
The 8 House is located in Ørestad South on the edge of a canal with a view of the open spaces of Kalvebrod Fælled in Copenhagen. With 475 units in a variety of sizes and layouts, the building meets the needs of people in all of life's stages: young and old; nuclear families and single people; families that grow and families that become smaller.

The bow-shaped building creates two distinct spaces, separated by the center of the bow which host the communal facilities of 5,300 sf. At the very same spot the building is penetrated by a 30-foot-wide passage that connects the two surrounding city spaces: the park area to the west and the channel area to the east. Instead of dividing the different functions of the building – for both habitation and trades – in separate blocks, they have been spread out horizontally.

The apartments are placed at the top, while the commercial program unfolds at the base of the building. As a result the apartments benefit from sunlight, fresh air and

Engineer: Moe & Brødsgaard
Owner: Høpfner A/S, Danish Oil Company, St. Frederikslund
工程师：莫伊&布洛德斯加尔德
所有人：霍普佛纳公司、丹麦石油公司、圣·弗莱德里克斯兰德

Architect / 建筑师
BIG
BIG

Location / 项目地点
Copenhagen, Denmark
丹麦，哥本哈根

Photo Credit / 图片版权
© Dragor Luftfoto, © Jens Lindhe, © Julien Lanoo,
© Jan Magasanik, © Ty Stange, © Ulrik Reeh
德拉格尔·卢福特弗托、詹斯·林德、朱利安·拉诺、詹·马加萨尼克、泰·斯坦格、乌尔里奇·利

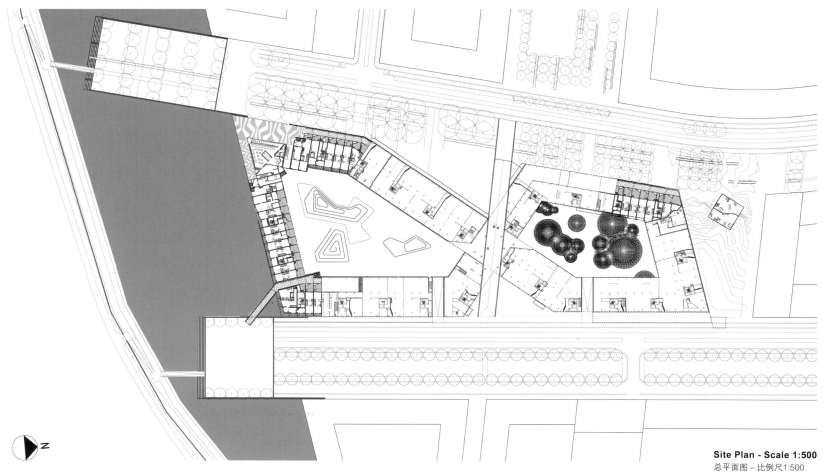

Site Plan - Scale 1:500
总平面图 – 比例尺1:500

the view, while the commercial spaces merge with life on the street. The 8 House has two sloping green roofs totaling over 1,700 m², which are strategically placed to reduce the urban heat island effect as well as to visually tie it back to the adjacent farmlands towards the south.

The shape of the building allows for daylighting and natural ventilation for all units. In addition, rainwater is collected and repurposed through a stormwater management system.

项目特色

8字住宅位于南奥莱斯塔德的一条运河边上，享有哥本哈根广阔的视野。住宅拥有475套不同规模、不同布局的公寓，能够满足不同年龄层次人群的需求——无论他们是年轻还是年老、小家庭或是单身人士、扩张的家族还是缩小的家族。

弓形建筑营造了两个独特的空间，二者被弓形结构的中央的公共区域隔开。9米宽的走廊连接了两个周边的城市空间：西面的公园和东面的海峡。项目没有将建筑的两个功能区——住宅区和商业区分设在两个楼内，而是以水平方向依次展开。公寓被设在顶部，而商业设施则设在建筑下半部分。因此，公寓享受了良好的阳光、新鲜空气和视野，而商业空间则融入了街面生活之中。8字住宅的两个斜坡绿色屋顶总面积超过1,700平方米，有效地减少了城市热岛效应，并且让建筑与南面的农田联系起来。

建筑的造型保证了各个公寓的自然采光和自然通风。此外，雨水被收集起来，通过雨水处理系统得到了重新利用。

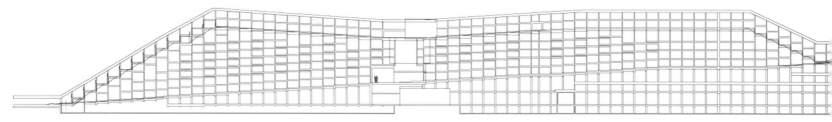

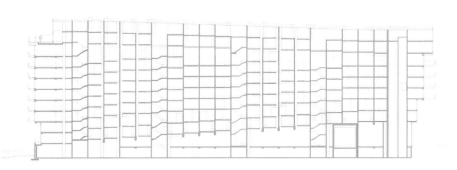

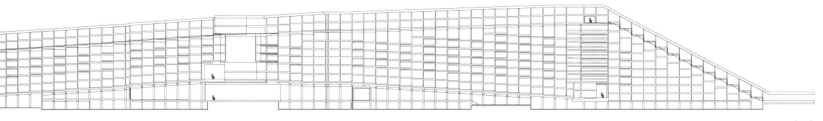

313

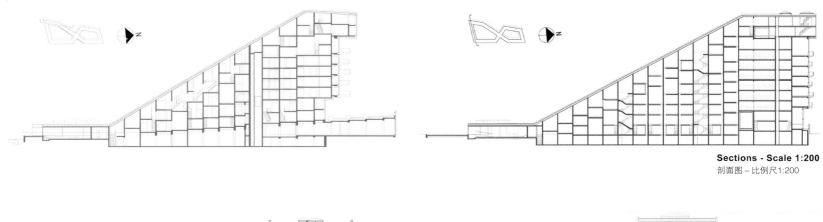

Sections - Scale 1:200
剖面图 – 比例尺1:200

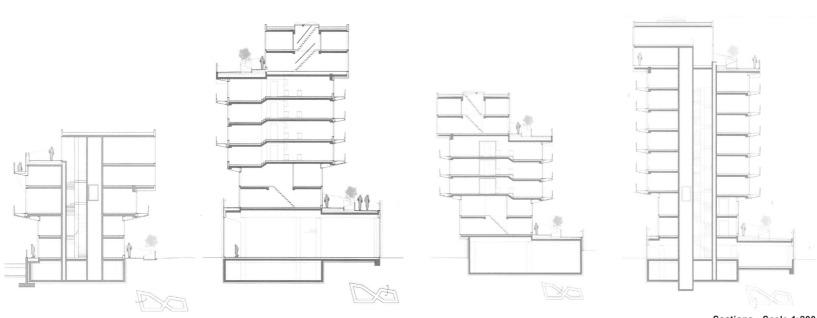

Sections - Scale 1:200
剖面图 – 比例尺1:200

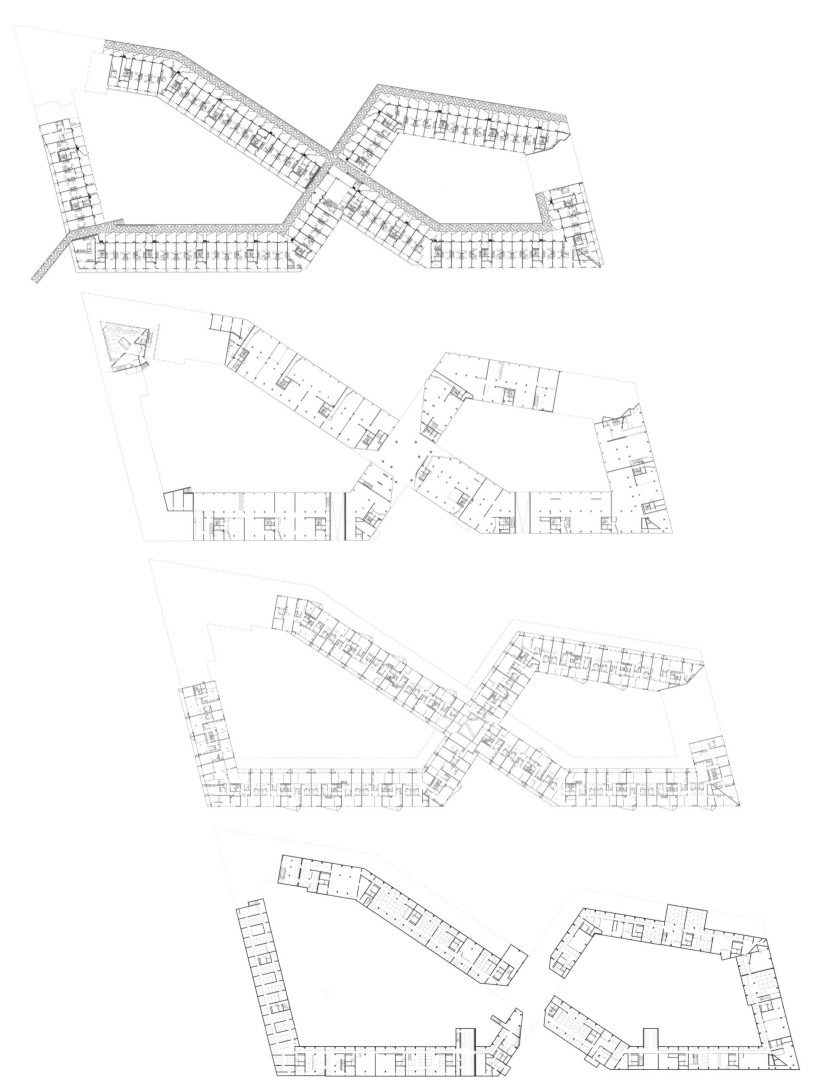

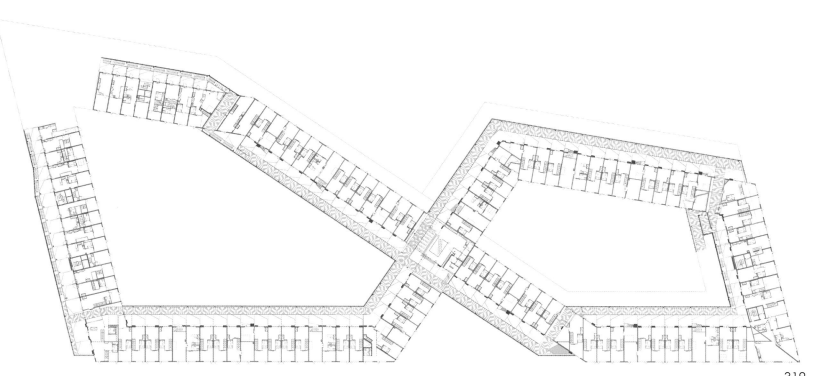

319

41 Cooper Square
库伯广场41号

Jury Comments:
A lot of attention was paid to the façade and the treatment of the interiors and it shows in how this project frames the timelessness with the intention of the building.
41 Cooper Square is so optimistic and wonderfully celebratory as you move through it. This has a spirit and aura to it that's extremely hard to capture and goes beyond most buildings.

评委评语：
建筑师对外立面和室内环境的处理耗费了大量心力，项目兼具功能性和永恒的价值。
行走于建筑内部，库伯广场41号的设计是如此令人欢乐而愉悦。
项目拥有一种难以捉摸的精神和气氛，超越了大多数建筑。

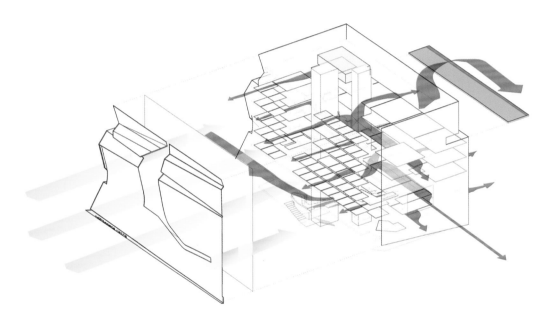

Notes of Interest
41 Cooper Square, the new academic building for The Cooper Union in New York City, aspires to manifest the character, culture, and vibrancy of both the 150 year-old institution and of the city in which it was founded. The institution remains committed to Peter Cooper's radically optimistic intention to provide an education "as free as water and air" and has subsequently grown to become a renowned intellectual and cultural center for the City of New York.

41 Cooper Square aspires to reflect the institution's stated goal to create an iconic building – one that reflects its values and aspirations as a center for advanced and innovative education in Art, Architecture and Engineering. Internally, the building is conceived as a vehicle to foster collaboration and cross-disciplinary dialogue among the college's three schools, previously housed in separate buildings.

A vertical piazza – the central space for informal social, intellectual, and creative exchange – forms the heart of the new academic building. An undulating lattice envelopes a 20-foot-wide grand stair which ascends four stories from the ground level through the sky-lit central atrium, which itself reaches to the full height of the building.

This vertical piazza is the social heart of the building, providing a place for impromptu and planned meetings, student gatherings, lectures, and for the intellectual debate that defines the academic environment.

Associate Architect: Gruzen Samton, LLP
Lighting: Horton Lees Brogden Lighting Design, Inc
Landscape Architect: Signe Nielsen
Acoustics: Newson Brown Acoustic, LLC
Owner: The Cooper Union
合作建筑师： 古鲁森·萨姆顿公司
照明设计： 霍顿·李斯·布罗格登照明设计公司
景观建筑师： 西格纳·尼尔森
声学： Newson Brown Acoustic, LLC
所有人： 库伯联盟学院

Architect / 建筑师
Morphosis Architects
墨菲西斯建筑事务所

Location / 项目地点
New York City, New York
纽约州，纽约

Photo Credit / 图片版权
© Iwan Baan
伊万·班

项目特色

库伯广场41号是库伯联盟学院的新教学楼。目标是展示这座具有150年历史的学院和纽约城的品质、文化和活力。学院一直秉承彼得·库伯的积极态度,提供"自由教育",逐渐成长为纽约市的著名学术文化中心。

库伯广场41号力求打造一座标志性建筑——反映学院的价值以及成为艺术、建筑和工程创意教育中心的渴望。从学院内部来看,建筑是促进学院内三个学院相互交流的工具。

垂直广场——用于非正式社交、教学和交流的中央空间——形成了新教学楼的中心。波浪形格架包围着一个6米宽的楼梯,从一楼中庭向上延伸了四层楼,直达建筑顶部。

这个垂直广场是学校的社交中心,提供了用于临时和计划性会面、学生集会、讲座以及学术讨论等空间。

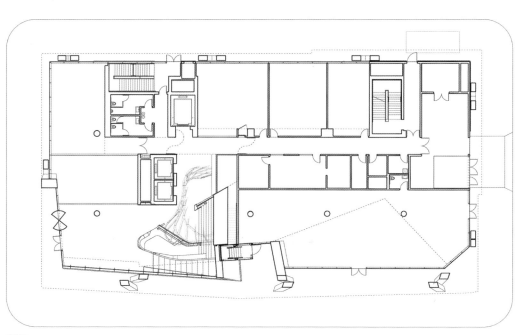

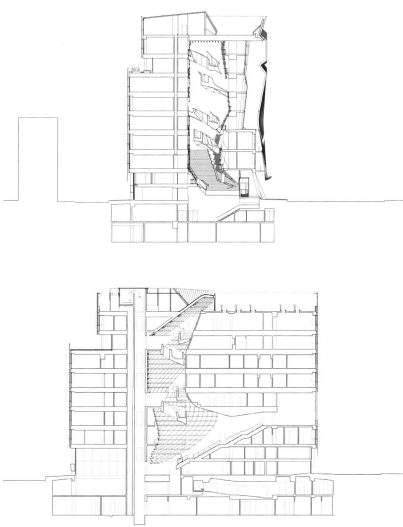

323

The Gates and Hillman Centers for Computer Science

盖茨和希尔曼计算机科学中心

Jury Comments:
This project is scaled perfectly within an urban campus and within a uniquely difficult site. The building not only matches the culture and aspirations of the school but also provides campus connections that had been clearly missing before. The fenestration and zinc exterior skin surprisingly relate beautifully to the campus fabric without being literal. Perhaps the most wonderful aspect of the project is a set of views and visual connections created by transparent interior glazing, non-reflective exterior glazing as well as carefully placed and angled floor plates.

评委评语：
作为一座地处城市校园和复杂场地的建筑，项目的比例规划十分完美。建筑不仅反映了学校的文化和启发性，而且完善了校园的连接性。门窗布局和锌制外壳为校园带来了毫不夸张的美感。
也许项目最重要的部分是它通过透明的室内玻璃、非反射性的外部玻璃记忆巧妙地布置楼面而形成了良好的视野感和连接性。

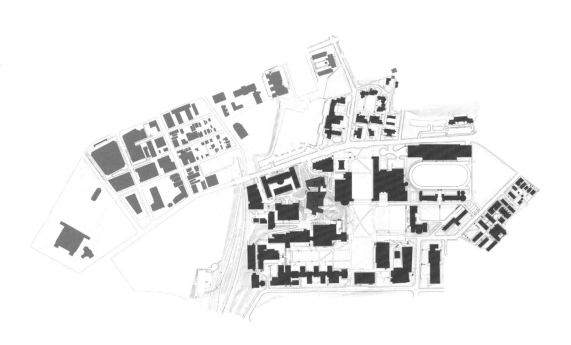

Notes of Interest

The Gates and Hillman Centers for Computer Science completes a computer science complex on Carnegie Mellon University's west campus. The building houses four departments of the School of Computer Science providing offices, conference rooms, open collaborative spaces, closed project rooms and a reading room for more than 120 faculty, 350 graduate students, 100 researchers or postdoctoral fellows and 50 administrative staff members along with a more public component of 10 university classrooms, a 250-seat auditorium, a Café and 2 university computer clusters.

The design of the Gates and Hillman Centers required negotiating a series of complex existing site conditions and programmatic pre-requisites. Site challenges included demolition of existing buildings, a large zone of subsurface rock, existing sewer lines that limited the constructable area, and an existing campus spacial hierarchy that had to be respected.

Programmatic pre-requisites included the need for a single building that could be treated as two separate buildings, the need for a variety of campus connections, both for pedestrians and for service purposes across a terrain that included variations of up to 75 feet in elevation.

Associate Architect: Gensler
Local Architect: EDGE Studio
Owner: Carnegie Mellon University
合作建筑师：詹斯勒
当地建筑师：EDGE 工作室
所有人：卡耐基梅隆大学

Architect / 建筑师	Location / 项目地点	Photo Credit / 图片版权
Mack Scogin Merrill Elam Architects MSME 建筑事务所	Pittsburgh, Pennsylvania 宾夕法尼亚州，匹兹堡	© Timothy Hursley, © Nic Lehoux 狄默思·赫斯利、尼克·卢克斯

项目特色

盖茨和希尔曼计算机科学中心为卡耐基梅隆大学的西校区带来了一座计算机科学大楼。大楼内设有计算机科学学院的四个部门,为超过120名教职员工、350名研究生、100名研究员和博士后学者以及50名行政员工提供了办公室、会议室、开放式合作空间、封闭项目室和阅览室,以及更开放的空间——10间教室、一间可容纳250人的礼堂、一间咖啡厅和两间计算机教室。

盖茨和希尔曼计算机科学中心的设计解决了一系列复杂的场地环境和规划要求。场地问题包括:拆除原有建筑、处理大面积的地表岩石、处理限制建设的污水管道以及尊重现有的校园空间层级。

项目的规划要求包括:打造一座可一分为二的单体建筑、与校园进行良好的连接以及横跨一个地势差高达23米的地面。

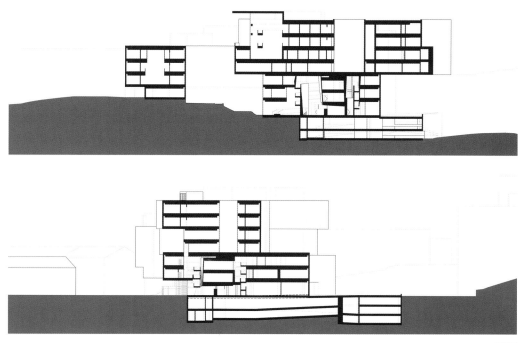

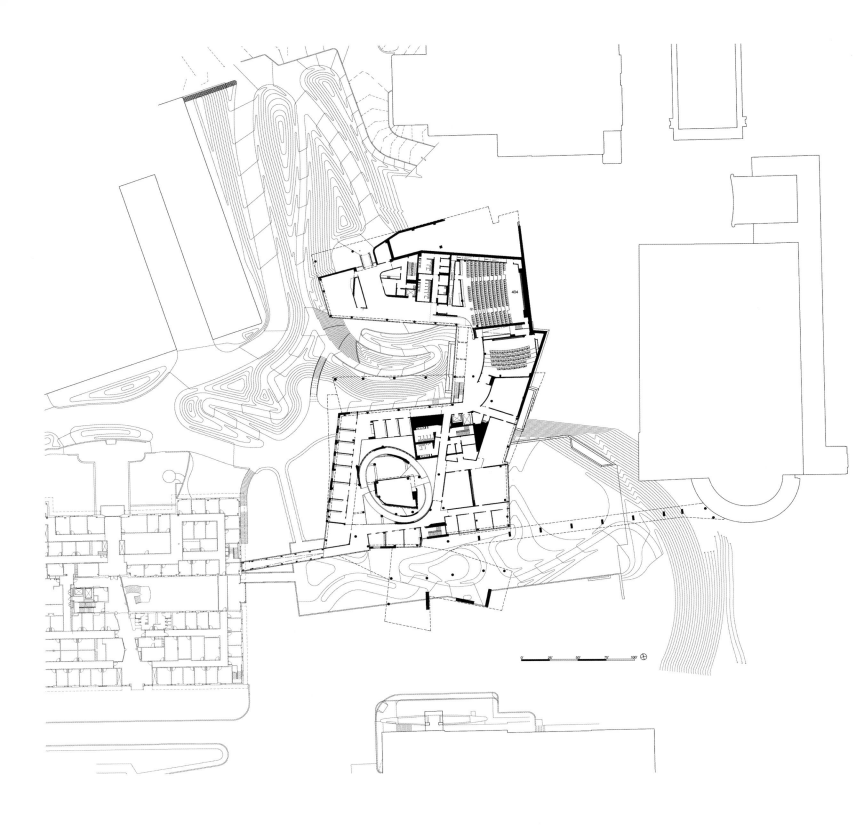

© Timothy Hursley

© Nic Lehoux

Ghost Architectural Laboratory
幽灵建筑实验室

Jury Comments:
This project reveals itself as more than just a grouping of buildings; it is a physical experiment in education as well as an act of will to preserve the serene beauty in the landscape. As a teaching tool the students find themselves immersed in an environment where they are challenged to produce high quality designs they can self construct.
This project is truly more than the sum of its parts; it is a wonderful resolution of materials, details, landscape, and learning.

评委评语：
项目不仅仅是建筑的结合；它既是教学的试验场，又以行动保护了景观。作为一种教学工具，学生们沉浸其中，不断进行高质量的设计。
项目真的不只是零件的汇总；它完美地结合了材料、细部、景观和教学。

Engineer: Campbell Comeau Engineering Ltd.
Building: Gordon MacLean Builders
工程师：坎贝尔·科莫工程公司
建造：戈登·麦克林建筑公司

Notes of Interest
The Ghost Laboratory is sited at the LaHave River estuary on Nova Scotia's Atlantic coast, where Samuel de Champlain made his first landfall in the new world in 1604. This landscape was re-cleared from forest by the architect over the past 25 years, revealing its historic ruins and its 400 years of agrarian history.
The Ghost Lab is an architectural education center in the tradition of Frank Lloyd Wright's Taliesin or Samuel Mockbee's Rural Studio. The permanent structures which now occupy the site among the ruins – tower, studio, cabins, barns and boathouse – are, in part, products of the design/build curriculum itself. They provide accommodation for the program and a venue for community events.
Each component started as a two-week project; from design, to foundation, to framing, to sheathing. The tower, which marks the south corner of the courtyard, and the barn, are sited just outside the fence and are built on wood post foundations. The studio and four cabins inside the fence are heated structures on concrete foundations. Each of the cabins is a prototypical and modest, 700-square-foot, two-bedroom structure comprised of a "servant" box and a "served" shed, clad in eastern cedar shingles. The 90 foot long, metal clad studio is dominated by a 40-foot worktable and a 72-foot totemic cedar cabinet. The 72-foot Ghost 9 barn contains an equipment shed and free stalls for horses, while creating a second working courtyard. The historic octagonal Troop barn, which has been moved 200 miles and redesigned to fit its new home, contains a community gathering hall on top and sheep stables below.

项目特色
幽灵建筑实验室位于新斯科舍省大西洋海岸线上的拉哈沃河口——萨缪尔·香普兰于1604年在此发现新大陆。最近25年来，建筑师重新清理了这里的森林，再现了它的历史遗迹和400年的耕地史。
幽灵实验室是一间建筑教育中心，就像弗兰克·洛依德·赖特的塔里埃辛和萨缪尔·莫克比的乡村工作室一样。这些永久性结构将与历史遗迹共存在场地上。塔楼、工作室、小屋、马厩、船库都将是设计/建筑课程的一部分。它们为各种项目和集体活动提供了容身之处。
每个元件都始于为期两周的项目；从设计、奠基、架构到外壳结构。庭院南角的塔楼和马厩设在围墙之外，采用木桩地基建造。围墙内的工作室和四间小屋采用了混凝土地基。每座小屋的面积都为65平方米，由"服务"和"被服务"两部分组成，外部覆盖着东部雪松木板。27米长的金属外壳工作室内拥有一张12米长的工作台和一个22米长的雪松橱柜。22米长的幽灵9号马厩内设有一间设备库和独立的马厩，同时也打造了第二工作庭院。古老的八角形军队马厩被移动了320公里，经过了重新设计，形成了社区集会大厅（上）和羊圈（下）的混合体。

Architect / 建筑师
Mackay-Lyons Sweetapple Architects Limited
幽灵建筑实验室

Location / 项目地点
Upper Kingsburg, Nova Scotia
加拿大新斯科舍省，上金斯伯格

Photo Credit / 图片版权
© Manuel Schnell, © Brian MacKay-Lyons, © James Steeves
曼纽尔·施奈尔、布莱恩·麦基—莱昂斯、詹姆斯·斯蒂夫斯

LumenHAUS
流明屋

Jury Comments:
The creative use of materials and the flexibility of its components quickly respond to changes in the environment through automated systems that optimize energy consumption. The plan and section are orchestrated by light and materials to enhance the perception of a small footprint.
The interior is cleverly designed with comfortable if compact spaces, compatible materials, and a rationale and clear layout.

评委评语：
材料的创意运用和元件的灵活运用能够通过优化能源消耗的自动系统快速应对环境变化。
平面和横截面通过光线和材料协调在一起，将小型建筑面积扩大化。
室内设计舒适而温馨，结合了紧致空间、共存材料和清晰的布局。

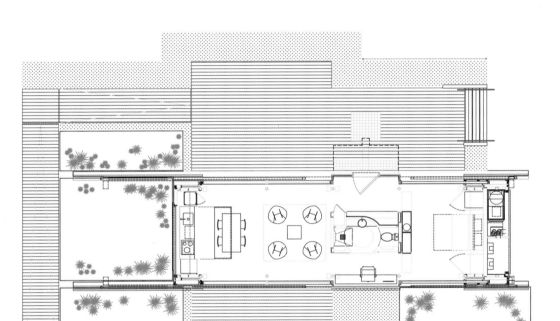

Notes of Interest
The house is both a dwelling and an exhibition informing the public about issues of alternative energy and sustainability. It has been exhibited in Washington D.C., Times Square, Madrid, Spain, Millennium Park, and at the Farnsworth House in Plano Ill.
This is a grid-tied solar powered house based on the concept of "Responsive Architecture". It adjusts to climactic changes and user requirements through automated systems that optimize energy consumption while offering an architecture of delight.
As a net-zero energy house employing active and passive systems, it generates more power than it uses over the course of a year. It achieves its positive energy balance by incorporating a contemporary reinterpretation of the architectural shutter and screen with innovative technology.
Built and operated using industrialized processes, the Eclipsis System optimizes energy use, makes building more efficient, and improves the quality of architectural space. The Eclipsis System is made of two exterior layers – laser cut stainless steel shutter screens and aerogel filled polycarbonate insulation panels – both of which integrate the house's technical and architectural identities. Rich and divergent qualities of light fill the house from sunrise to sunset, and sliding panel systems automatically respond to climactic conditions, providing a full range of protection from the elements and a rich architectural experience.

Structural Engineer: ARUP
Cladding/Material Fabrication: Zahner and Associates, Inc.
Control Systems: Siemens
Geothermal Materials: Mechanical Equipment Sales
Hardware: Hafele America Co.
Photovoltaics: Solar Connexions; Baseline Solar, AltEnergy, SMB Solar, RTKL
Owner: School of Architecture + Design, Virginia Tech
结构工程师： ARUP
包层/材料装配： 扎娜事务所
控制系统： 西门子
地热材料： 机械设备销售公司
计算机硬件： 美国海福乐公司
光电能设备： 太阳连接公司、基线太阳能、阿尔特能源、SMB太阳能、RTKL
所有人： 弗吉尼亚理工学院建筑设计学院

Architect / 建筑师
Center for Design Research, School of Architecture + Design, Virginia Tech
弗吉尼亚理工学院建筑设计学院设计研究中心

Location / 项目地点
Blacksburg, Virginia
弗吉尼亚州，布莱克伯格

Photo Credit / 图片版权
© Virginia Tech Solar Team
弗吉尼亚理工学院学者团队

项目特色

流明屋既是住宅又是展览,像公众们展示了替代能源和可持续设计方案。项目在华盛顿、时代广场、马德里、千禧公园以及普拉诺的范斯沃斯住宅进行了广泛的展览。

这是一座以"敏感建筑"为主题的太阳能动力住宅。它通过自动化系统来响应气候变化和用户需求,在优化能源消耗的同时打造了令人愉悦的建筑。

作为采用主动与被动系统的一座零能耗住宅,它的年能源生产量大于年能源消耗量。通过使用创新工艺的百叶窗和遮阳板实现了积极地能源平衡。

省略系统利用工业化进程建造和操作,优化了能源利用,提高了建筑的效率和建筑空间的质量。省略系统由两层结构组成——激光切割的不锈钢遮阳板和气凝胶树脂隔热板——二者统一了建筑工艺和建筑特性。丰富而分散的光线从早到晚一直充满住宅,滑动面板系统则自动应对气候变化,提供了全方位的保护和丰富的建筑体验。

Pittman Dowell Residence
皮特曼·道威尔住宅

Jury Comments:
The house acts like an optical instrument with staged views of the surrounding landscape including spectacular views of the valley below and the hills above.
The clear concise presentation of details and the theatrically arranged spaces constitute a sublime and poetic expression and push the boundaries of what a house can be.

评委评语:
住宅像一个光学仪器一样,展现了周边山谷的壮丽景色。
细部和空间布局的清晰呈现组成了一种庄重而诗意的表达方式,突破了住宅建筑的界限。

Notes of Interest
North of Los Angeles at the edge of Angeles National Forest, the residence is sited on 6 acres originally planned as a Richard Neutra designed development. Although three pads were cleared, only one home was built. The current owners, who own the original Neutra home, have developed a desert garden on one of the clearings. The new home sits on the last clearing.

Five decades after the original house was constructed, the site's visual and physical context has changed dramatically. Similarly, the contemporary needs of the artist residents required a new relationship between building and landscape.

Inspired by geometric arrangements of interlocking polygons, the home is a heptagonal figure whose purity is confounded by a series of intersecting slices. Bounded by an introverted exterior, living spaces unfold in a moiré of shifting perspectival frames. Movement and visual relationships expand and contract to respond to the centrifugal nature of the site and context. An irregularly shaped void defined by these intersections creates an outdoor room whose edges blur into the adjoining spaces.

Engineer: B.W. Smith Structural Engineers, Paller-Roberts Engineering, Inc., The J Byer Group
General Contractor: Asterisk Builders
Arborist: Robert W. Wallace
Owner: Lari Pittman and Roy Dowell
工程师: B·W·史密斯结构工程公司、帕勒尔-罗伯斯工程公司、J·拜尔集团
总承包商: 星号建筑公司
树木景观: 罗伯特·W·华莱士
所有人: 拉里·皮特曼和罗伊·道威尔

Architect /建筑师
Michael Maltzan Architecture, Inc.
迈克尔·马尔赞建筑设计公司

Location /项目地点
La Crescenta, California
加利福尼亚州,拉克里森塔

Photo Credit /图片版权
© Iwan Baan
伊万·班

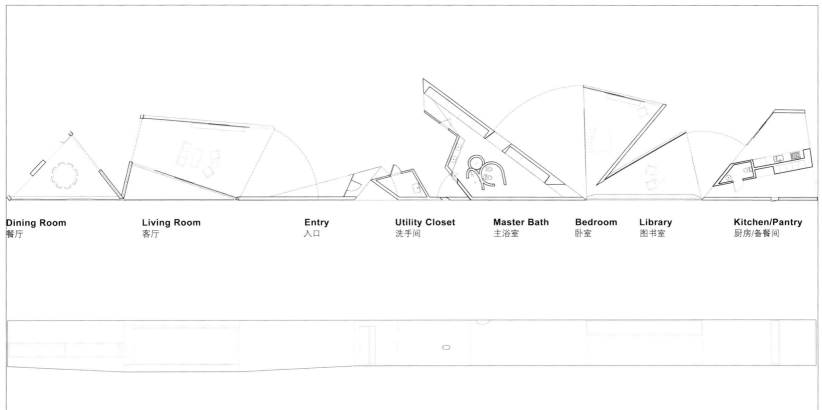

Dining Room 餐厅　　**Living Room** 客厅　　**Entry** 入口　　**Utility Closet** 洗手间　　**Master Bath** 主浴室　　**Bedroom** 卧室　　**Library** 图书室　　**Kitchen/Pantry** 厨房/备餐间

Structural Grid 结构网格

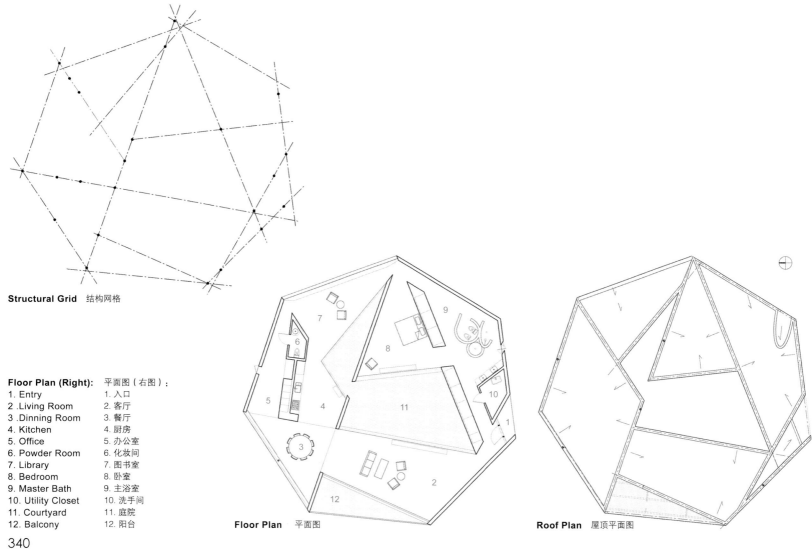

Floor Plan (Right): 平面图（右图）:
1. Entry — 1. 入口
2. Living Room — 2. 客厅
3. Dinning Room — 3. 餐厅
4. Kitchen — 4. 厨房
5. Office — 5. 办公室
6. Powder Room — 6. 化妆间
7. Library — 7. 图书室
8. Bedroom — 8. 卧室
9. Master Bath — 9. 主浴室
10. Utility Closet — 10. 洗手间
11. Courtyard — 11. 庭院
12. Balcony — 12. 阳台

Floor Plan 平面图

Roof Plan 屋顶平面图

项目特色

住宅位于洛杉矶北部安吉利斯国家森林的边缘，场地原来是理查德·纽特拉设计的房产开发项目。尽管已经清理了三块场地，最终却只建造了一座住宅。现在的所有人（拥有纽特拉住宅）在其中的一块空地上开发了一座沙漠花园。新住宅将建造在最后的一块空地上。

在原始住宅建造了50年之后，场地的视觉和物理环境经过了剧烈的变化。同样的，艺术家住户的现代需求也要求在建筑和景观之间建立全新的联系。

住宅设计从连锁的多边形结构中获得了灵感，采用了交叉切片装饰的七边形结构。起居空间被内翻的外壳包围，呈波纹框架展开。运动与直觉联系根据场地和环境的特性而不断伸缩。这些连锁空间之间所形成的不规则形状的空隙形成了边界模糊的户外空间。

Poetry Foundation

诗歌基金会

Jury Comments:
This building unfolds as it is experienced and is sublime in its stillness and detailing. From the street, one is seduced by is secrecy and upon entering its crafted inner court, the project is revealed much like a poetry reading.
The manipulation of light through sectional explorations and the weaving of its limited material use through its interiors are resolved exceptionally well.

评委评语:
建筑纯熟地铺展开来,在沉静和细节中达到了顶峰。
街道上的行人被它的隐密而吸引,一走进精雕细琢的内庭,项目便如诗歌般展开。
光线的控制和材料的组合让室内空间异常出色。

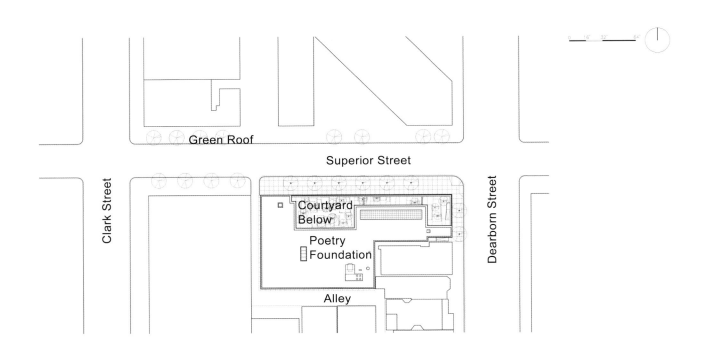

Notes of Interest

The Poetry Foundation is comprised of a building in dialogue with a garden created through erosion of an implied volume as described by the L-shaped property boundary. The garden interlocks with the building and is conceived as another "room", part of the building's slowly-unfolding spatial sequence revealed space by space, not unlike a poem is revealed line by line.

Visitors enter through the garden, an urban sanctuary that mediates between the street and enclosed building. Entering the garden, visitors first encounter the library space, announcing that they are entering into a literary environment. Inside, an exhibition gallery connects the library to the performance space, where visitors listen to poets read their work against the backdrop of the garden.

Public functions (performance space, gallery and library) are located on the ground floor, while office spaces are located on the second level, organized into three areas (Foundation Administration, Poetry magazine/website, and Programs). The building is configured to allow for views from all spaces out onto the garden.

Tectonically, the building is conceived of as a series of layers that visitors move through and between. Layers, of zinc, glass, and wood, peel apart to define the various spaces of the building. The building's outer layer of oxidized zinc becomes perforated where it borders the garden, allowing visual access to the garden from the street to encourage public investigation.

Engineer: ARUP; dbHMS; Terra Engineering
Development: U.S. Equities Realty; Norcon
Landscape Architect: Reed Hilderbrand Associates
Lighting: Charter Sills
Acoustical: Threshold Acoustics
Owner: Poetry Foundation
工程师:ARUP;dbHMS;特拉工程公司
开发商:美国股市不动产;诺尔康
景观建筑师:里德·希德尔布兰德事务所
灯光设计:查特尔·希尔斯
音响设计:门槛音效公司
所有人:俄亥俄州立大学

Architect / 建筑师
John Ronan Architects
约翰·罗南建筑事务所

Location / 项目地点
Chicago, Illinois
伊利诺伊州，芝加哥

Photo Credit / 图片版权
© Hedrich Blessing
赫德瑞奇·布莱辛

项目特色

诗歌基金会由一座大楼和一座花园组成，整个项目呈L形。花园与大楼相互交错，构成了另一个"空间"。建筑的一部分空间序列缓缓展开，宛如诗歌一行行展开。

游客们通过花园进入。花园是连接街道与封闭建筑的都市桃源。一走进花园，首先映入眼帘的就是图书空间，宣告着他们进入了一个文学环境。内部的展览厅连接着图书馆和表演空间，诗人们在花园的背景下朗诵诗歌。

公共功能区（表演空间、展览厅和图书馆）设在一层，而办公空间则设在二层，分成三个区域——基金会办公区、诗歌杂志/网站工作区和项目规划区。建筑的造型让各个区域都能看到花园的景色。

从构造上讲，建筑由一系列的层次组成，而访客们则游走于其间。锌、玻璃、木材等层次依次界定着建筑内的不同空间。建筑的外层采用氧化锌板，在花园处则呈现出开口，让街道上的行人也能享受花园的美景。

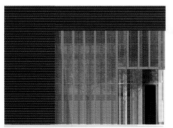

East Elevation　东立面

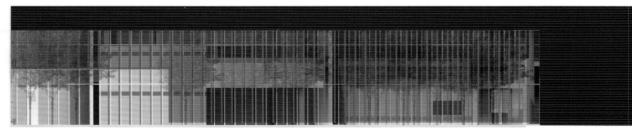

North Elevation　北立面

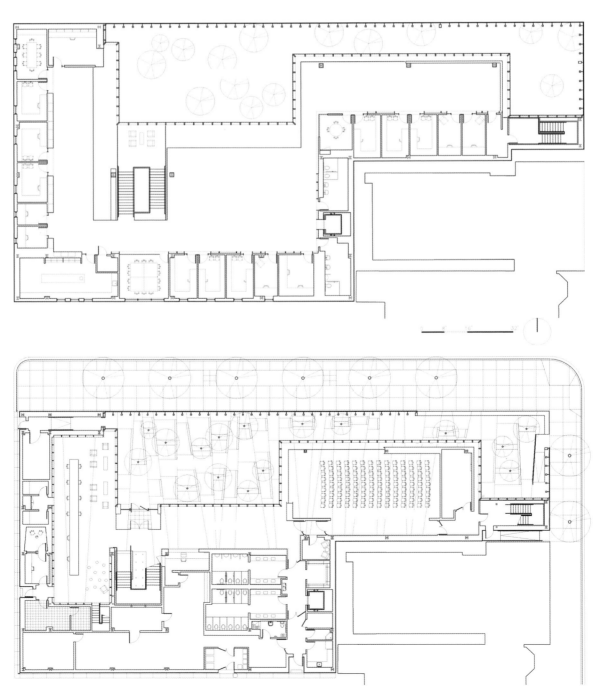

Ruth Lilly Visitors Pavilion
卢斯莉莉游客亭

Jury Comments:
This pavilion is artfully cast in the shadows of the adjacent trees, its transparency is enhanced by its latticed canopy which filters light thru its entirety and the floor to ceiling glazing hides no secrets.
Its low posture and horizontal form enhances the encompassing flora and is quite elegant in its lightness while reaching out and inviting nature in.

评委评语:
游客亭环绕在四周的树荫之中,网格穹顶进一步增添了它的通透感。日光从穹顶洒入亭内,而落地玻璃更是让内部一览无余。
它低矮的水平纵深造型突出了四周环绕的植物,而透明感则显得异常优雅,与自然融为一体。

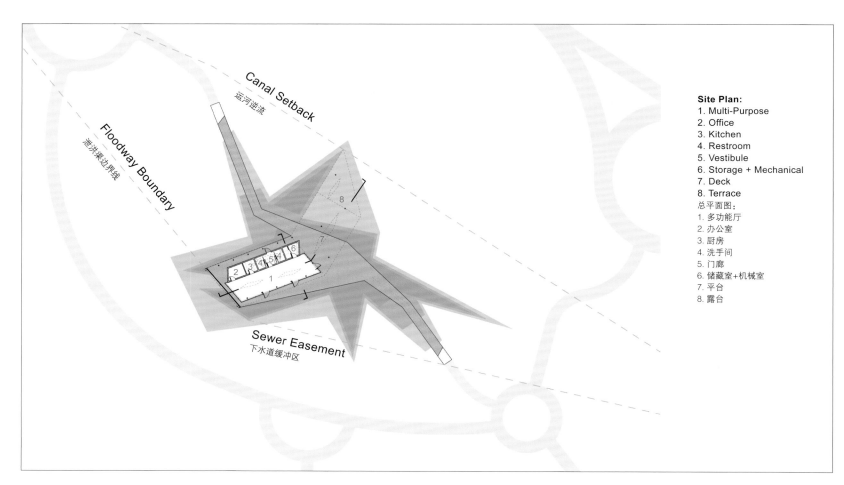

Site Plan:
1. Multi-Purpose
2. Office
3. Kitchen
4. Restroom
5. Vestibule
6. Storage + Mechanical
7. Deck
8. Terrace

总平面图:
1. 多功能厅
2. 办公室
3. 厨房
4. 洗手间
5. 门廊
6. 储藏室+机械室
7. 平台
8. 露台

Notes of Interest

The Ruth Lilly Visitors Pavilion is the result of a studied relationship between building, land and art, and serves as both a threshold to and a destination within the 100 Acres Art & Nature Park at the Indianapolis Museum of Art.

The goal was to craft a logic of ideas and physical works that reveal the repressed raw power of environment, art, and architecture through a rethinking and re-making of "where we already are". Questions are born from within the found condition, resulting in an immanent response, imbued with conviction, meaning, and significance for the Museum of Art, its patrons and the citizens of Indianapolis.

The Visitors Pavilion is a place of shared resolve where nature and artifice are sensually perceived as one and many; the detail and horizon. The 100 acre park site is born of wildly turbulent natural and cultural phenomena constantly changing the land's structure, and is a place where one becomes conscious of the residual forms that reveal the creative life force at work in our world.

Tinkering with it as cultivated urban wilds proves a sound means of joining Nature and City as a recovered, unpredictably changing but cultivated landscape. Prone to flooding across the entire park by the White River, the park offers less than an acre for the construction of the Ruth Lily Visitors Pavilion.

Engineer: Guy Nordenson & Associates, L'Acquis Consulting Engineers, Cripe Architects & Engineers
General Contractor: Geupel DeMars Hagerman
Landscape Architect: The Landscape Studio; NINebark
Owner: Indianapolis Museum of Art

工程师:盖·诺尔登森事务所、拉奎斯工程咨询公司、克里普建筑工程公司
总承包商:GDH
景观建筑师:景观工作室、九层皮公司
所有人:印第安纳波利斯艺术博物馆

Architect / 建筑师	Location / 项目地点	Photo Credit / 图片版权
Marlon Blackwell Architect 马龙·布莱克威尔建筑事务所	Indianapolis, Indiana 印第安纳州，印第安纳波利斯	© Timothy Hursley 狄默思·赫斯利

项目特色

卢斯莉莉游客亭的设计研究了建筑、地景与艺术的关系，是印第安纳波利斯艺术博物馆艺术与自然公园的门户，也是其中的一个景点。

目标是打造具有逻辑的观念和实体，通过重新思考人类自身的位置来展示环境、艺术及建筑一直被压抑的原始力量。现有条件引出了许多问题，最终形成了内在的响应，为项目注入了艺术博物馆、赞助人和印第安纳波利斯市民的信念、意义及重要性。

游客亭是人们分享自然与技巧的关系的场所，充满了细部设计与水平设计。占地100英亩的公园由狂野的大自然和现代文明组成，不断变换着地面的结构，人们在这里意识到了展示创意生活力量的残余形式。

项目与城市荒野的融合将自然与城市结合成一个复原而文明的景观。在遭受怀特河的洪水侵袭之后，公园仅余下1英亩的场地来建设卢斯莉莉游客亭。

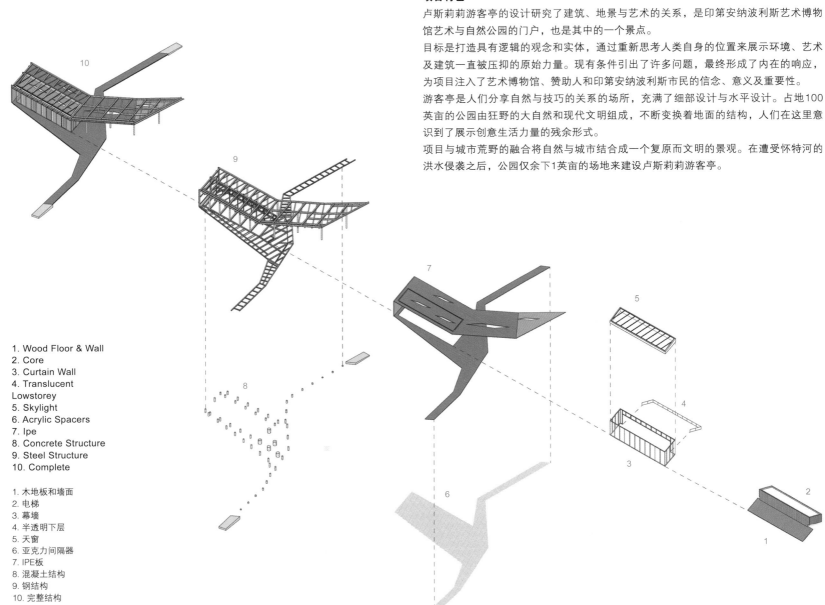

1. Wood Floor & Wall
2. Core
3. Curtain Wall
4. Translucent Lowstorey
5. Skylight
6. Acrylic Spacers
7. Ipe
8. Concrete Structure
9. Steel Structure
10. Complete

1. 木地板和墙面
2. 电梯
3. 幕墙
4. 半透明下层
5. 天窗
6. 亚克力间隔器
7. IPE板
8. 混凝土结构
9. 钢结构
10. 完整结构

The Standard, New York
纽约标准酒店

Jury Comments:
The building addresses the urban scale as a tower relating to highline and river well. It blends seamlessly with the fabric of the surrounding neighborhood.
There is clarity in the choice and articulation of materials and a sense of restraint, though the end result is one of high visual impact.
The goal for transparency and openness successfully drives design and detailing decisions.

评委评语:
建筑在城市中以高塔的形式呈现,与高架桥和河流进行了良好的联系。它与周边的建筑网格完美地结合在一起。
材料的选择十分明晰,既具有约束感,又达成了醒目的视觉效果。
该项目设计和细节的选择完美地达成了项目的通透感和开敞感目标。

Notes of Interest
Located in Manhattan's Hudson riverfront Meatpacking District, the hotel responds to its context through contrast: sculptural piers, whose forms clearly separate the building from the orthogonal street grid, raise the building fifty-seven feet off the street, and allow the horizontally-scaled industrial landscape to pass beneath it and natural light to penetrate to the street.
The 18-story building straddles the High Line, a 75-year-old elevated railroad line recently developed into a new linear, public park. The two slabs of the building are "hinged", angled to further emphasize the building's distinction from the city's grid and its levitation above the neighborhood.
The low-scale environment affords the building unique visibility from all directions, and unobstructed 360° views of the city. The juxtaposition of the building's two materials – concrete and glass – reflects the character of the city: the gritty quality of the concrete contrasts with the refinement of the glass. The concrete grid provides a delicate frame for the exceedingly transparent water-white glass, the two materials unified in the continuous plane of the curtain wall.
The curtain wall breaks with the traditional architecture of hotels, replacing opacity with transparency, privacy with openness and defining a new paradigm.

Engineer: DeSimone; Edwards & Zuck, H.A. Bader, Langan Engineering & Environmental Services
Façade: R.A. Heintges
Interior Design: André Balazs Properties; Shawn Hausman; Roman and Williams
Acoustical: Cerami & Associates
Owner: André Balazs Properties
工程师:德塞蒙、爱德华兹&扎克、H.A.巴德尔、兰甘工程环境服务公司
外立面设计:R.A.海因特吉斯
室内设计:安德烈·巴拉斯地产、肖恩·霍斯曼、罗曼和威廉姆斯
音响设计:瑟拉米事务所
所有人:安德烈·巴拉斯地产

Architect / 建筑师	**Location** / 项目地点	**Photo Credit** / 图片版权
Ennead Architects 恩尼德建筑事务所	New York City, New York 纽约州，纽约	© Alex MacLean/Landslides Aerial Photography © Jeff Goldberg/Esto, © Aislinn Weidele/Ennead Architects 艾利克斯·麦克林/兰斯里德斯航空摄影、杰夫·格德伯格/Esto、艾斯林·威德尔/恩尼德建筑事务所

© Jeff Goldberg/Esto

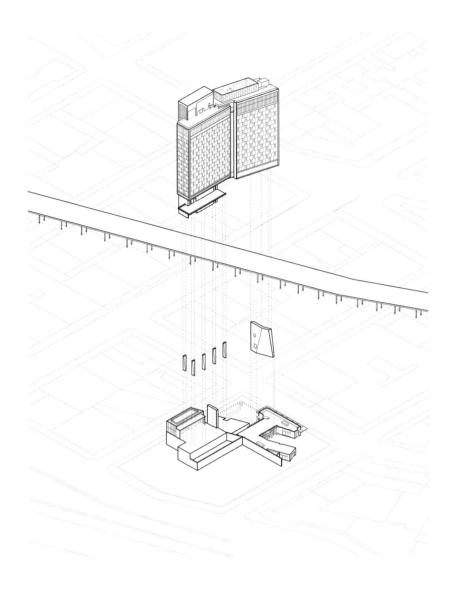

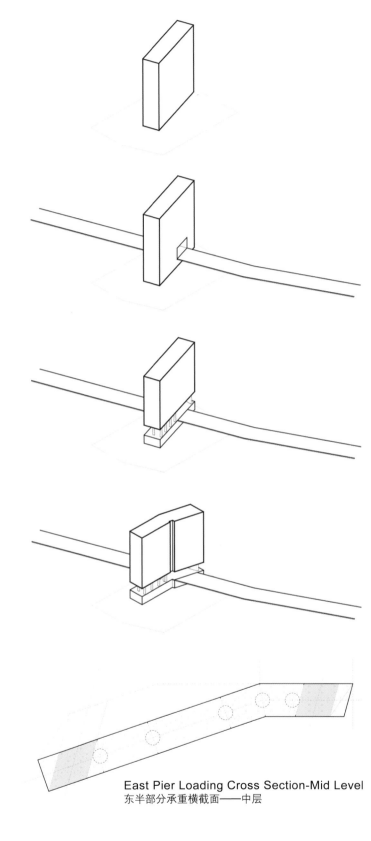

Massing Transformation　体块变形

East Pier Geonmetry　建筑东半部分

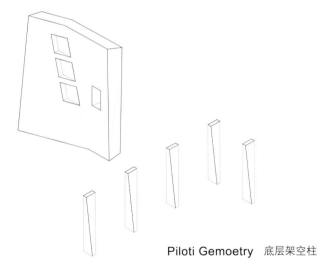

Piloti Gemoetry　底层架空柱

East Pier Loading Cross Section-Mid Level
东半部分承重横截面——中层

East Pier Loading Cross Section-Lower Level
东半部分承重横截面——下层

项目特色

酒店位于曼哈顿休斯顿河畔的肉类加工区，通过对比与周边的环境形成了互动：雕塑支柱将建筑与直角形的街区隔开，使它脱离地面，让下方的工业景观水平穿过，也让阳光洒在街面上。

18层高的建筑跨越了高架铁路带状公园。建筑的两面呈铰链形，进一步凸显了它在城市网格中的与众不同。

四周低矮的建筑环境让酒店大楼在各个方向都享有良好的视野，无障碍观看360度城市全景。建筑材料——混凝土和玻璃——的并置反映了城市的特色：坚忍不拔的混凝土与精致的玻璃形成了对比。混凝土为透明如水的玻璃提供了精妙的框架，两种材料在幕墙上形成了统一的整体。

幕墙打破了传统式酒店设计，重新混合了透明与不透明、私密与开放，形成了全新的典范。

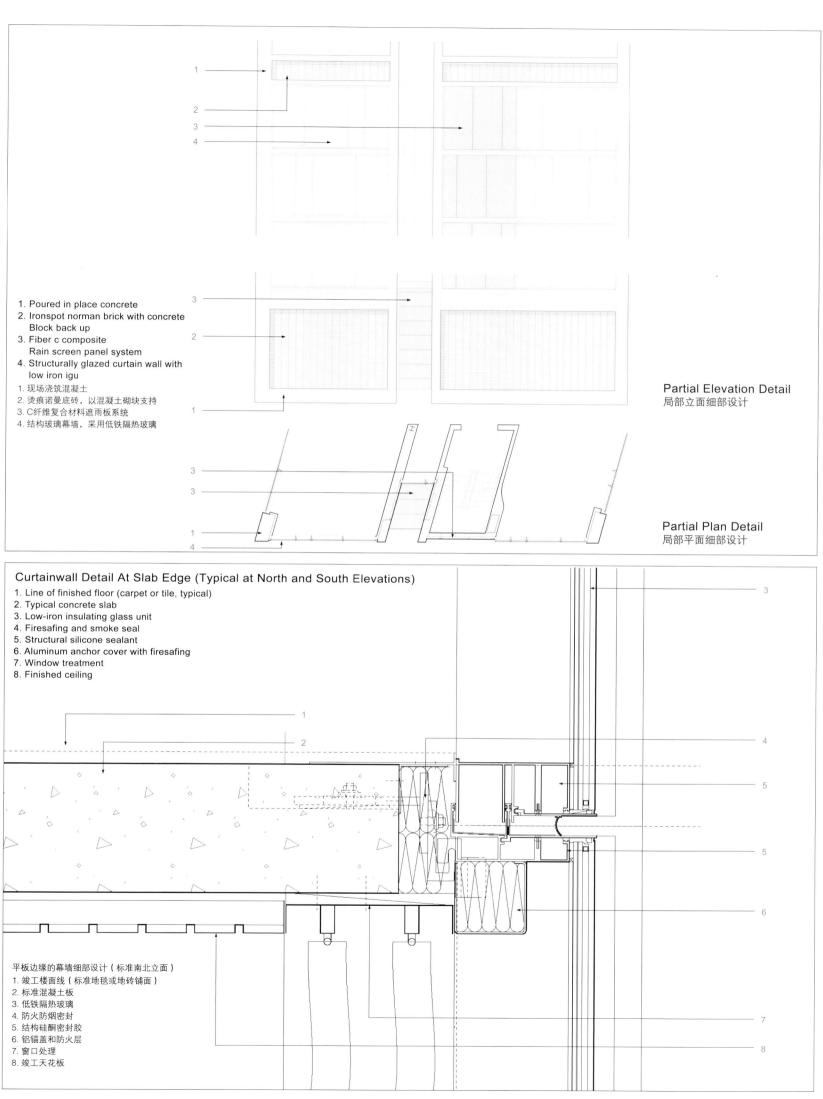

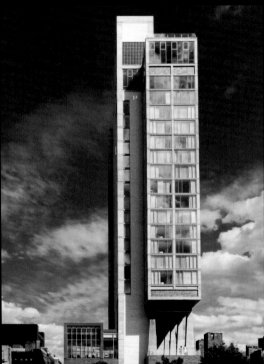

© Jeff Goldberg/Esto

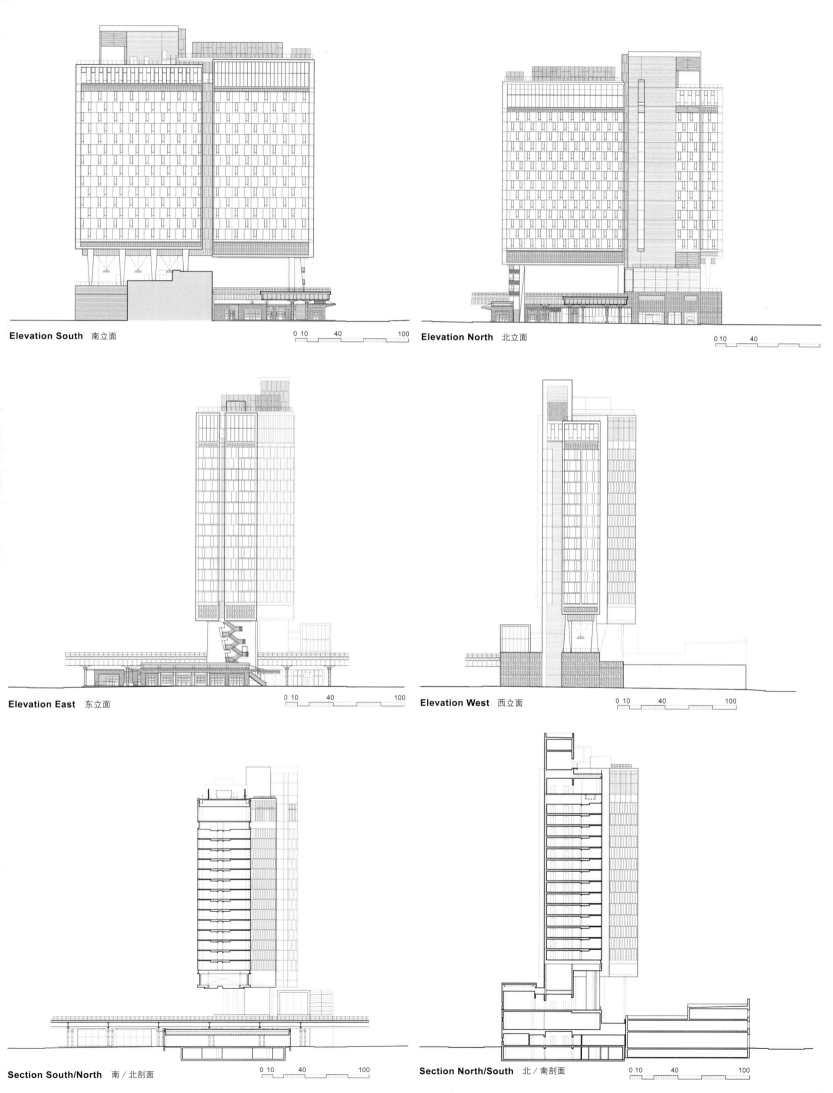

© Aislinn Weidele/Ennead Architects

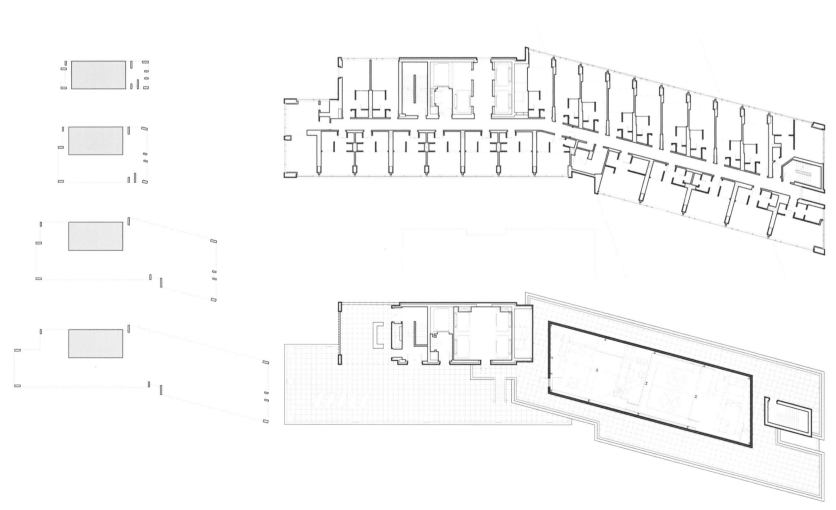

Level 1 一层

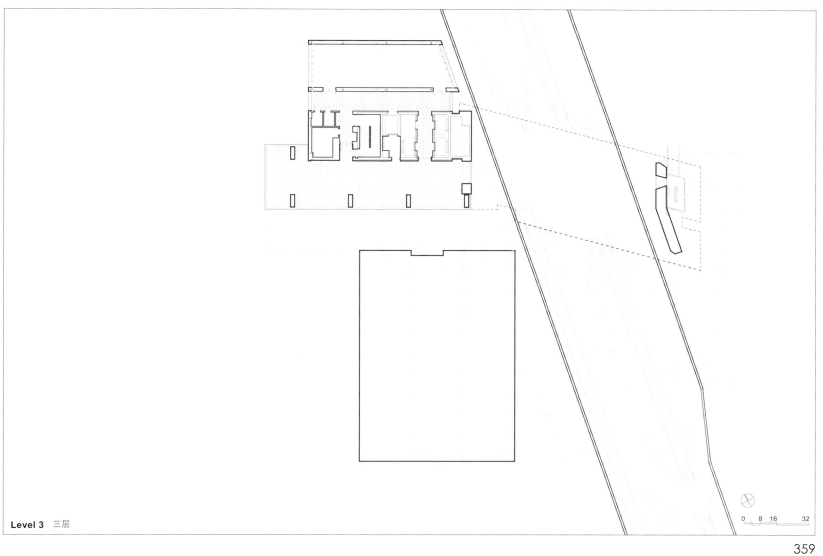

Level 3 三层

ARTifacts
工艺工作室

Jury Comments:
An excellent example of what is possible within limited means and unlimited desire. Working within a historical structure, the project sensitively responded to both program and context in a series of deliberate and carefully considered moves, keeping with the spirit of the place and its inspiration.
The solution was both raw and refined.

评委评语：
项目在有限的方式和无限的欲望之间形成了非凡的表达。
项目利用历史结构对项目和环境进行了回应，做出了一系列精心考虑的举动，保留了空间的精神和灵气。
设计方案质朴又精致。

Notes of Interest

Kent Bellows (1949-2005) was a lifelong Nebraska Artist who has been recognized as one of America's greatest masters of American Realism and was known as a mentor for friends, colleagues and burgeoning artists.

The Kent Bellows Foundation requested a renovation to the artist's work/live building to transform it into a center for art and student mentoring. The Kent Bellows Studio and Center for Visual Arts strives to ignite the creative spark in inner city youth, encouraging them to reach their highest potential through self expression in the visual arts. Artists ages 12-21 attend workshops and receive mentoring from nationally renowned artists through various programs and events put on by the Center.

The artist had worked in the building for twenty years and had built out much of the space with his own hands. Their intention was to identify the artifacts the artist left that had meaning. They identified and preserved 9 artifacts (gallery floor, Kent's parka, glass block transom, moving backdrop wall, books, collages, light fixture pulley, wall mural, and wall installation).

Working with the artifacts, the design focused on minimal interventions to upgrade the building and provide the new spaces for the facility. The storefront intervention was a three dimensional sculpture of steel plates/tubes which creates windows, seating, facility signage, and the main entrance. The staircase/balcony intervention creates a continuous steel plate walkway that connects the entrance, gallery, library, office and the second floor studios. The library intervention is a meeting and reading space hovering above the gallery defined by a folded wood panel wall/ceiling that frames the artist's moving backdrop wall.

Construction: University of Nebraska Architecture Students
Owner: Kent Bellows Studio and Center for Visual Arts
施工： 内布拉斯加大学建筑系学生
所有人： 肯特·贝劳斯工作室和视觉艺术中心

Architect /建筑师
Randy Brown Architects
兰迪·布朗建筑事务所

Location /项目地点
Omaha, Nebraska
内步拉斯加州，奥马哈市

Photo Credit /图片版权
© Farshid Assassi
法尔希德·阿萨西

Plan Key :
1. Student Work Space
2. Gallery
3. Bellows' Preserved Studio
4. Library
5. Office
6. Bellows' Preserved Mural Space

设计重点：
1. 学生工作区
2. 走廊
3. 贝劳斯的工作室
4. 图书室
5. 办公室
6. 贝劳斯的壁画区

Basement 地下室   First Floor 一层   Mezzanine 中夹层   Second Floor 二层

项目特色

来自内布拉斯加州的艺术家肯特·贝劳斯（1949–2005）是美国最伟大的美国现实主义艺术大师，他是同行和年轻艺术家的良师益友。

肯特·贝劳斯基金会要求将一座艺术家的工作/生活楼改造为艺术和学生指导中心。肯特·贝劳斯工作室和视觉艺术中心力求点燃城市年轻人的创造灵感，鼓励他们通过在视觉艺术中的自我表现来达到自己的潜力的顶峰。

肯特曾经在这座楼里工作了20年，楼内的大部分空间都由他亲手打造。我们的意图是突出他所留下的具有意义的物品。我们坚定并保留了9件物品——画廊地板、肯特的派克大衣、玻璃砖气窗、移动背景墙、书籍、剪贴画、灯具滑轮、墙画和墙面装置。

设计力求通过最小限度的改动来对建筑升级并打造新空间。店面的三维雕塑由铁板和铁管组成，创造出窗户、座椅、引导标示和主入口。楼梯/阳台的改造打造了一个连续的钢板走道，连接了入口、画廊、图书室、办公室和二楼的工作室。图书室被改造成一个会议和阅读空间，以折叠的木板墙/天花板为特色，将肯特的移动背景墙框了起来。

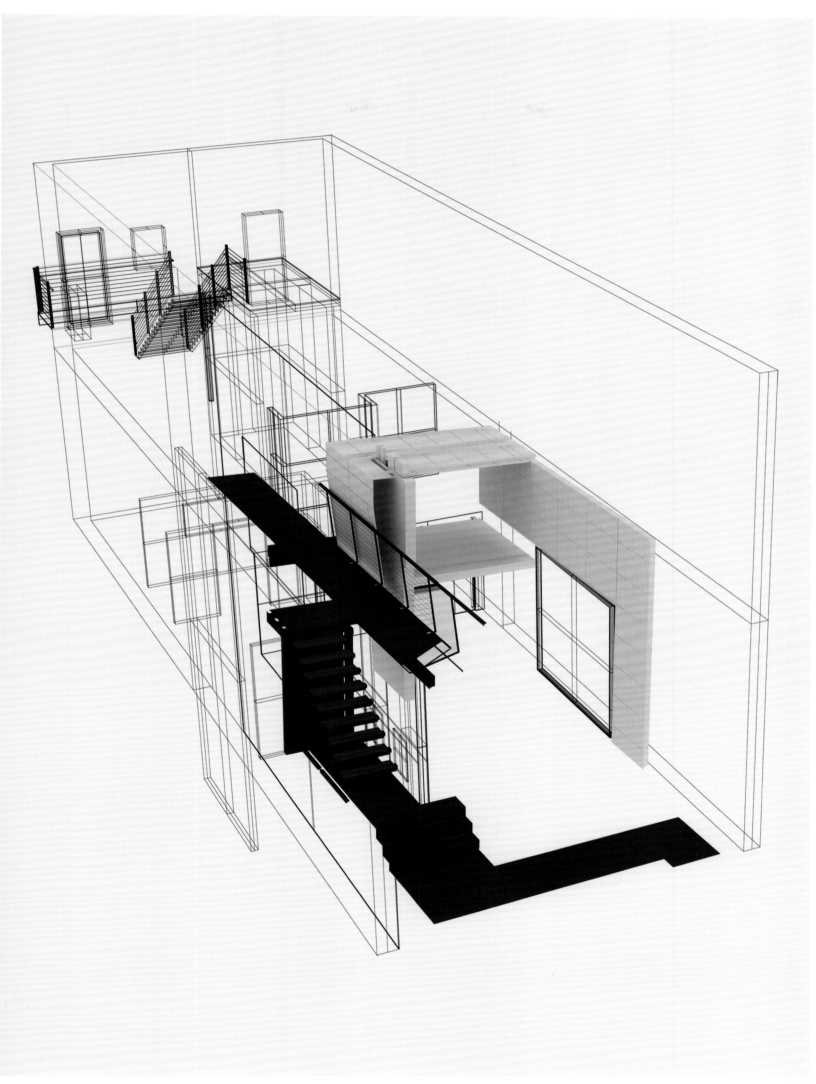

Children's Institute, Inc. Otis Booth Campus
儿童学院公司奥的斯园区

Jury Comments:
An excellent example of adaptive re-use, this project creates a safe place for families exposed to violence.
The spatial zone between the north portion and the southern portion is well considered, effectively serving to open the building to natural light and interactive activities.

评委评语：
作为建筑再利用的典范，项目为遭受暴力的家庭打造了安全的空间。
项目南北两侧的空间区域经过了精心考量，有效地为建筑引入了自然光线并提供了互动空间。

Engineer: John A. Martin & Associates, Inc., KMA Consulting, KPFF Consulting Engineers
General Contractor: Swinerton
Environmental Graphics: Newsom Design
Geotechnical: Geotechnologies, Inc.
Hardware: Finish Hardware Technology
Landscaping: Nancy Goslee Power & Associates
Lighting: Horton Lees Brogden Lighting Design
LEED: AECOM
Signage: Newsom Design
Specifications: Specifications West, LLC
Waterproofing: IRC Waterproofing
Owner: Children's Institute, Inc.

工程师：约翰·A·马丁事务所；KMA咨询公司；KPFF工程咨询公司
总承包商：思维诺顿
环境图形：纽萨姆设计
土工技术：土工技术公司
硬件：装修五金技术公司
景观设计：南希·高丝里·波尔事务所
照明设计：霍顿·里斯·布罗格顿照明设计
绿色建筑相关：AECOM
引导标示：纽萨姆设计
性能规范：西部性能规范公司
防水：IRC防水
所有人：儿童学院公司

Notes of Interest
The adaptive reuse of three industrial buildings created the headquarters for a non-profit organization that assists children and families exposed to violence.
The campus is split by an alley with the north site focusing on preschool and early childhood services and the south site anchored around a community center offering educational programs (art, technology, nutrition, and after-school) as well as counseling services.
Therapy rooms are dispersed around community spaces to make visits an everyday, rather than clinical, experience. This innovative strategy deinstitutionalizes the services' traditional delivery and builds trust in a neighborhood in need of both counseling and community programs.
On a tight budget ($10.5 million), the design provides required amenities that also add a sense of identity and welcome. A key part of the process was re-thinking program organization to reveal opportunities for creative and collaborative community engagement.

Architect / 建筑师	Location / 项目地点	Photo Credit / 图片版权
Koning Eizenberg Architecture 科宁·艾森伯格建筑事务所	Los Angeles, California 加利福尼亚州，洛杉矶	© Eric Staudenmaier 埃里克·斯托登梅尔

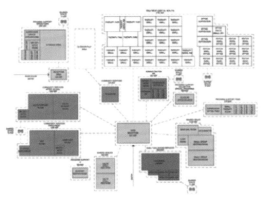
1. Owner's Program 1. 业主规划

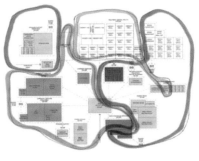
2. Re-grouping 2. 重置规划

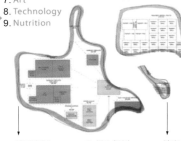

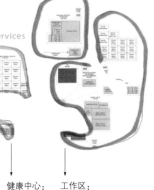

Community Space
1. Preschool
2. Teen center
3. Family resource center
4. Events
5. Drop-in programs
6. Training
7. Art
8. Technology
9. Nutrition

Wellness
Therapy
Support services

Workspace
Team based administrative space

社区空间：
1. 学前区
2. 青少年中心
3. 家庭资源中心
4. 活动室
5. 插入规划
6. 培训室
7. 艺术室
8. 技术室
9. 营养室

健康中心：
治疗室
附属服务

工作区：
团队式行政空间

项目特色

三座工业楼被重新改造成为一家非营利机构的总部，该机构旨在帮助遭受暴力的儿童和家庭。

园区被一条小路一分为二：北侧以学前和低龄儿童服务为主，南侧则环绕着一个社区中心展开，主要提供教育项目服务（艺术、技术、营养和课外活动）和咨询服务。

心理治疗室围绕在社区空间四周，营造出不同于诊所的日常氛围。这一创新方式打破了心理治疗的传统方式，在需要咨询和社区项目的附近街区之间建立了信任。

项目的预算很紧（1,005万美元），既提供了必要的设施，又增添了辨识度和温馨感。设计流程最重要的部分就是重新思考功能区的组织，展现创新型社区参与。

3. Social Setting 3. 社会环境

From Program to Social Setting
从规划到社会环境

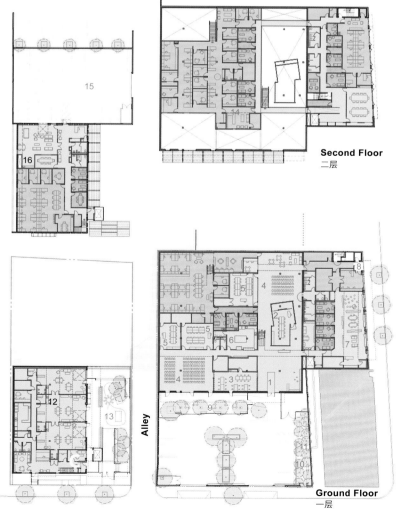

Key: 主要标示：

▢ Workplace 工作区
▢ Wellness Program 健康项目
▢ Community Space 社区空间

Second Floor 二层

Ground Floor 一层

1. Entry
2. Curiosity Box Around Tech Center
3. Art Room
4. Big Room/Space
5. Activity Room
6. Café/Nutrition
7. Family Resource Center
8. Deliveries
9. Patio/Garden
10. Orange Grove
11. Edible Garden
12. Preschool
13. Play Yard
14. Workplace Reception
15. Half Court/Parking
16. Teen Center

1. 入口
2. 技术中心四周的空间
3. 艺术室
4. 大活动室
5. 活动室
6. 咖啡厅/营养室
7. 家庭资源室
8. 交货室
9. 天井/花园
10. 橘子林
11. 蔬果园
12. 学前区
13. 游戏场
14. 工作区前台
15. 半边庭院/停车场
16. 青少年中心

David Rubenstein Atrium at Lincoln Center

林肯中心大卫·鲁宾斯坦中庭

Jury Comments:
This space is sensitively transformed incorporating nature, art, and commercial activity in a carefully modulated manner. The multiple scales created by the ceiling "puddles" and huge custom textile art are brought down to human scale by the inclusion of mural plantings, tables and chairs, reading areas, and well situated vendor stations.
This is an exceptional revitalization of an urban interior space that is both uplifting and considerate.

评委评语:
这个空间以精致的方式结合了自然、艺术与商业活动。天花板上的"漩涡"和巨大的定制纺织艺术品为中庭打造了多重空间结构,并且通过壁画、桌椅、阅览区和贩售摊的设计而回到了人性尺度。
这是一个非凡的城市室内空间,既振奋人心又精致体贴。

Notes of Interest

Harmony atrium, a privately owned public space, was a defacto homeless shelter and small rock-climbing business. Lincoln Center sponsored the space with the true intention of creating a place for the public.

Wedged into Manhattan's dense fabric, the 7,000 sf passageway serves as Lincoln Center's public visitor facility, welcoming city newcomers and neighborhood residents. The space, known as The David Rubenstein Atrium at Lincoln Center, offers free performances, information and tickets to events, and a place to have a cup of coffee or a glass of wine.

Cantilevered canopies announce the presence of the atrium. Visitors enter through large glass doors. They are greeted by 20-foot-high plant walls. Green marble benches, as well as moveable chairs and tables, offer places to rest. A fountain in the ceiling drops thin streams of water into a stone basin. Sixteen occuli pierce the golden ceiling to bring natural light into the double height space. In the evening, they are illuminated with colored artificial lights creating an ideal atmosphere for concerts.

Enormous felt paintings hang on two walls. One installation, grey ellipses rolling playfully on a yellow background, relates to the ceiling, and the other surrounds a media wall that serves as a canvas for projected information, images, and film. Transformed by light, color, texture, and thoughtfully chosen materials, the space is now a tranquil and welcoming oasis. In the first five months, more than 250,000 people visited the atrium.

Consultant: Acoustic Dimensions; Axis Group Limited; Dan Euser Waterarchitecture Inc; Fisher Dachs Associates; Pentagram Design, Inc.; Steven Winter Associate, Inc.; Vertical Garden Technology
Engineer: ARUP
General Contractor: RCDolner Construction LLC
Lighting: Fisher Marantz Stone
Owner: Lincoln Center for the Performing Arts

顾问: 声学维度公司、轴线集团、丹·厄斯尔水上建筑公司、费舍尔·达奇斯事务所、五星设计公司、史蒂文·温特事务所、垂直花园技术公司
工程师: ARUP
总承包商: RC多尔纳建筑公司
照明设计: 费舍尔·马兰士·斯通
所有人: 林肯表演艺术中心

Architect / 建筑师 Tod Williams Billie Tsien Architects TWBT建筑事务所	**Location** / 项目地点 New York City, New York 纽约州，纽约	**Photo Credit** / 图片版权 © Nic Lehoux, © Tod Williams Billie Tsien Architects 尼克·卢克斯、TWBT建筑事务所

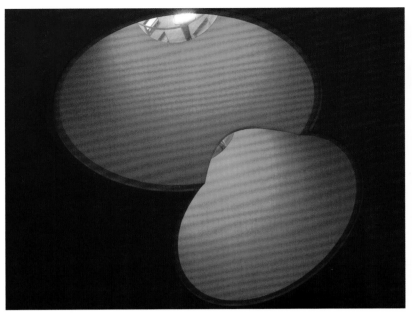

项目特色

和谐中庭是一个私有的公共空间，为无家可归的人提供庇所并经营小小的攀岩生意。林肯中心真心想为公众打造一个空间。

这个650平方米的通道嵌入了曼哈顿密集的城市网络，是林肯中心的公共访客设施，吸引着游客和附近的居民。大卫·鲁宾斯坦中庭提供免费的表演、信息和活动门票，同时也是小酌的好去处。

悬臂式穹顶是中庭的典型标志。人们通过巨大的玻璃门进入，首先映入眼帘的便是6米高的植物墙。绿色大理石长椅和移动桌椅提供了休息空间。从天花板上倾泻而下的喷泉直达石水池。天花板上的16个穿孔为室内带来了自然光线。晚上，五颜六色的灯光将会为演唱会营造理想的氛围。

中庭的两面墙壁上挂着巨大的毛毡绘画作品。天花板的黄色背景上点缀着灰色椭圆结构；而媒体墙后方的同类型装置像一块帆布，用于展示投影信息、图像和影片。空间通过光线、色彩、纹理和精选材料的装饰，成为了宁静而温馨的绿洲。在最初的五个月，有超过250,000人来到了中庭。

HyundaiCard Air Lounge
现代信用卡机场休息室

Jury Comments:
The project takes an innovative approach to the airline lounge model, effectively establishing a unique relationship between the passenger and the space.
The well-conceived assimilation of technology engages the traveler in both the "black box" and the surrounding walls that integrate the helpful passenger flight status flip-screens.
Extraordinarily clean detailing that does not come at the expense of function or friendliness.

评委评语：
项目创造性地打造了航线休息室模型，有效地在乘客和空间之间建立了独特的联系。
精心构思的技术融合不仅让旅客们在"黑盒子"可以充分休闲，也在四周墙壁上的翻转屏上输入了有助的航班信息。
非凡而简洁的细部设计并不会削弱该项目的功能价值和友好的气氛。

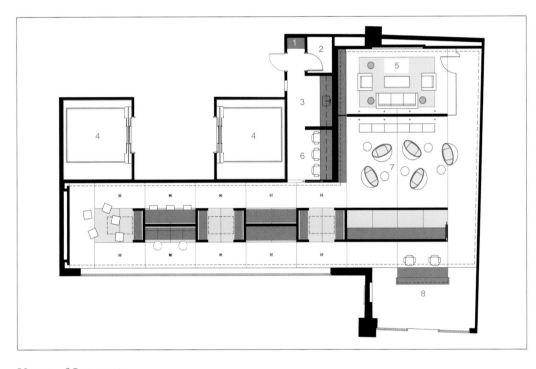

1. Locker 　　1. 更衣室
2. Janitor Closet 2. 守卫室
3. Pantry 　　3. 备餐间
4. Freight Elevator 4. 货梯
5. Small Vip 　5. 小型贵宾休息室
6. Office 　　6. 办公室
7. Large Vip 　7. 大型贵宾休息室
8. Reception 　8. 接待处

Consultant: Laschober + Sovitch
Engineer: Kesson International
Lighting: Kaplan Gehring McCarroll
Owner: HyundaiCard Company
顾问： 拉朔博尔+索维奇
工程师： 科松国际
灯光设计： 卡普兰·格林·麦克卡罗尔
所有人： 现代信用卡公司

Notes of Interest

Located at Incheon Airport in Korea, this 250 sm project seeks to create an exclusive environment to offer unique travel assistance for HyundaiCard Black members. As a counterpoint to the surrounding visual noise and frenzied airport activity, the proposed parti is deceptively simple, with functions arranged in a freestanding "black box".

Much like a perfectly organized suitcase, this monolithic object contains all the information, accessories, entertainment, and gifts needed for a memorable travel experience. The HyundaiCard space shifts the paradigm of a traditional lounge by combining lounge, retail, and museum programs. Rather than a static place for waiting, it is a dynamic space one passes through to better prepare for the trip ahead. As such, visitors are able to pace their movements through the space according to individual needs, desires, and schedules.

Among the unique features in the lounge are a custom vending machine, fantastic dream-like art movies by Hiraki Sawa, and a personalized flight tracking system. Also, there are two virtual skylights in the black box, both of which move slowly through the color spectrum of the sky. These spaces act as haven-like environments in which travelers become aware of the sky's variations, thus establishing a symbolic, if not poetic relationship with the notion of air travel.

Within the constraint of a small envelope, reflective surfaces provide visual relief while cove lighting plays up the ethereal atmosphere of the space.

项目特色

项目位于韩国仁川机场，总面积250平方米，试图为现代信用卡"黑卡"持有人提供独一无二的专属出行服务。与熙熙攘攘的机场氛围形成了对比，休息室简单清雅，各个功能区被安排在一个独立的"黑盒子"之中。

就像井然有序的手提箱一样，这个巨大的结构容纳着旅行所需的所有信息、设备、娱乐以及礼品。现代信用卡休息室改变了典型的传统休息室空间，集休息室、零售店和博物馆于一身。它不是一个静态的等候区，而是一个引领人们准备前方旅行的动感空间。这样一来，游客们能够根据个人需求、渴望和行程来安排自己的行动。

休息室的特色包括定制的自动贩卖机、泽拓所设计的梦幻艺术影片以及个性化航班跟踪系统。此外，黑盒结构上的两个虚拟天窗缓缓地变化呈现着天空的色彩。这些空间形成了天堂般的环境，旅客们能够清楚地知道天空的变化，与航空飞行形成了形象而诗意的联系。

在局促的小型外壳内，反光表面材料释放了空间，而凹圆暗槽灯则让空间变得缥缈而优雅。

Architect / 建筑师
Gensler
詹斯勒事务所

Location / 项目地点
Incheon, South Korea
韩国，仁川

Photo Credit / 图片版权
© Ryan Gobuty
赖安·谷布提

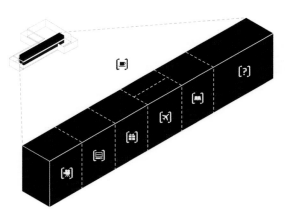

[?] Concierge 前台
[✈] Travel Accessories 旅行用品
[▦] Printed Materials 印刷资料
[▤] Food & Refreshments 食品
[▣] Giveaways 赠品
[▥] Business Center 商务中心
[▨] Entertainment 娱乐
[▩] VIP Lounge 贵宾休息室

How much time do you have?

· · · · 05 Minutes [?]
· · · · 15 Minutes [?]+[✈]
· · · · 30 Minutes [?]+[✈]+[▦]
· · · · 45 Minutes [?]+[▦]+[▨]+[▣]+[▥]
· · · · 60 Minutes [?]+[▦]+[▨]+[▣]+[▥]+[▩]

我们有多少时间?
05分钟
15分钟
30分钟
45分钟
60分钟

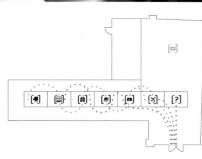

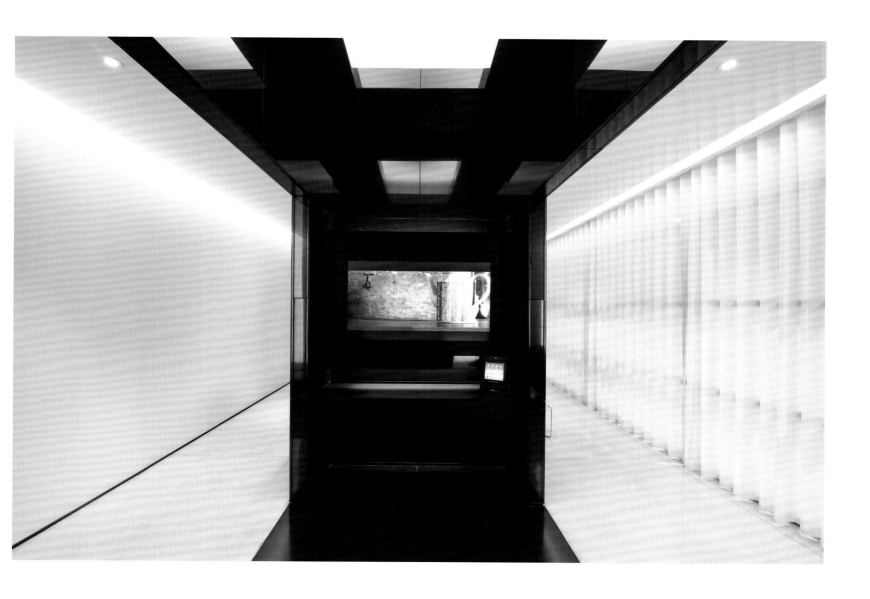

Virtual Skylights
The black boxes act as "skyspaces", i.e. impressively minimal structures with virtual skylights designed to change the way viewers perceive light. Because of the meditative way in which they channel light, the spaces act as haven-like environments in which travelers become aware of the sky's variations, thus establishing a symbolic, if not poetic, relationship with the notion of air travel.

虚拟天窗
黑盒子就好像是"天空"。例如，天花板上打造了虚拟天窗来反映外部光线的变化。由于它们会缓慢地调整光线，这些空间形成了天堂般的环境，旅客们能够清楚地知道天空的变化，与航空飞行形成了形象而诗意的联系。

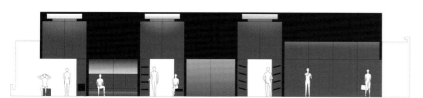

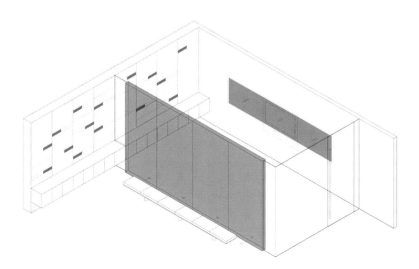

Flip Screens
Wanting to bring some of the excitement and romance associated with departures, we re-imagined the classic flip screen designed by Giorgio Segatto for the Solari di Udine Station in Milan. Upon check-in with the concierge, a traveler's information is relayed to one of fifteen LCD displays. Each LCD display cycles through a series of seven slides (guest name, airline, flight number, destination, gate/terminal, flight status and boarding time), thus creating a dynamic, personalized and artful way to ensure that card members make it on time.

翻转屏
为了给旅行带来一些振奋并浪漫的元素，我们重新改造了希奥尔希奥·西加图为米兰乌内迪车站设计的经典翻转屏。从前台登记进入休息室之后，旅客信息就被转存进了15个液晶显示器中的一个。液晶显示器以七行字幕进行滚动循环（旅客姓名、航线、航班号、目的地、登机口、航班信息和登机时间），由此形成了一种动态的个性化信息方式，保证了持卡人能准时登机。

The Integral House
整合住宅

Jury Comments:
The relationship of the home to both its musical program and its surrounding environment was superbly articulated.
The louvered vertical fins modulate the light and views to the exterior surroundings as well as correlate to music theory rooted in strong rhythm and syncopation. The fins added both measure and cadence to the overall movement.
The sensitivity, appropriate application, craft, and execution of detail were well executed.

评委评语:
住宅与音乐功能和周边环境的关系得到了良好的表达。
百叶窗式围栏调节了光线和视野,同时也形成了强烈的韵律感和停顿感。木围栏为项目的整体运动中增添了测量度和节奏感。
敏感度、适当的应用、工艺以及细部设计都得到了良好的执行。

Notes of Interest
The Integral House creates a place for architecture, music, and performance located at the threshold between Toronto's urban fabric and its extensive natural ravine system. In the project's program brief, the client clearly articulated his dual passion for mathematics and music and his interest in curvilinear shapes resulting in spatially complex volumes.

Viewed from its residential neighborhood, one reads a two story building with a grounded wood base sitting below a translucent gently shaped etched glass skin. The wooden base dissolves into oak clad fins echoing the undulating contour lines of the river valley and the winding pathways of the native forest of oaks, beaches, and maples. The main concert hall/performance space is located a full floor below your entry level and becomes intertwined with the verdant ravine landscape.

The project integrates many sustainable features into the site and building. A field of vertical geothermal pipes supplies heating and cooling for the entire project including the main concert hall/performance space for 150 – 200 people. A lush green roof is centrally located and a visual feature from many parts of the project. The vertical wooden fins provide sun shading from the exterior as well as contributing to the acoustical performance of the concert hall/performance space. Materials have been carefully selected for their aesthetic contribution as well as their enduring qualities based on life cycle costing calculations.

Engineer: Blackwell Bowick Partnership; DT Prohaska Engineering; Dynamic Designs and Engineering Inc.; Toews Engineering
Fountain Consultant: Waterarchitecture Inc.
General Contractor: Eisner Murray Custom Builders
Interior Design: Decisive Moment
Landscape Architect: NAK Design Group
Lighting: Suzanne Powadiuk Design Inc.
Acoustical: Swallow Consultants
Owner: Dr. James Stewart

工程师: 布莱克威尔·波维克事务所、DT普罗哈斯卡工程设计、动态设计和工程公司、突维斯工程
喷泉咨询: 水上建筑公司
总承包商: 艾斯纳·默里定制建造公司
室内设计: 决定时刻设计公司
景观建筑师: NAK设计集团
照明设计: 苏珊娜·波娃迪克设计公司
音效设计: 燕子咨询公司
所有人: 詹姆斯·斯图加特博士

Architect / 建筑师
Shim-Sutcliffe Architects
西姆–苏特克里夫建筑事务所

Location / 项目地点
Toronto, Canada
加拿大，多伦多

Photo Credit / 图片版权
© James Dow
詹姆斯·陶

项目特色

整合住宅在多伦多的城市网络和广阔的自然峡谷系统之间打造了一个融合了建筑、音乐以及表演艺术的空间。客户在项目要求中明确指出他对数学和音乐的热爱，以及对曲线造型的兴趣，以上元素综合形成了一个复杂的空间结构。

从周边的住宅看，整合住宅是一座以木材为底座、以半透明亚光玻璃为上半部分的两层高建筑。木制底座融入了橡木围栏之中，与河谷的曲线和森林中蜿蜒的小径遥相呼应。主要的音乐厅/表演空间设在入口楼层之下，与青翠的峡谷景观交织在一起。

项目综合了许多可持续特征。垂直地热管为整个项目（包括可容纳150-200人的音乐厅）提供了供暖和制冷。茂密的绿色屋顶设在中央，是项目的视觉特色之一。垂直的木围栏起到了遮阳作用并且有利于音乐厅的音响效果。精挑细选的材料既具美学价值，又具有持久的价值。

Joukowsky Institute for Archaeology & the Ancient World

儒科夫斯基考古和远古世纪学院

Jury Comments:
This is an extremely intelligent and compelling project on multiple levels.
The design makes a clever reference to its archaeological interests by creating a "found" object that is both beautifully detailed and sophisticated in expression.
The effort directed at dissolving the boundaries between student and teacher is admirable and the overall project renders a fresh, dynamic interior intervention that is both innovative and beautifully resolved.

评委评语：
从各个层面来讲，这都是一个非常明智且引人注目的项目。
设计巧妙地借鉴了考古学的价值，打造了兼具美观细部和精妙表述的"现存"空间。
用于解决学生与教师之间障碍的努力令人赞赏，整体项目做出了新鲜而动感十足的室内改造，既有创意，又有美感。

Notes of Interest

The Joukowsky Institute for Archaeology and the Ancient World completely reinvents Rhode Island Hall, a historic Greek Revival building at the center of the Brown University campus. It is an endowed institute, serving the Brown community with teaching, research, fieldwork, and classroom studies for both graduates and undergraduates.

This project restores Rhode Island Hall's exterior, and entirely renovates its interior. The extensive spatial and structural reconfiguration allows us to reconsider the ways that daylight was delivered throughout the project. Translucency of both glass and wood creates varying levels of transparency and daylight between program spaces, encouraging a more interactive dialogue between faculty and student.

The contemporary intervention within this historic shell challenges the notion of archaeology as a conservative and dusty pursuit, and supports the mission of the Joukowsky Institute as a progressive leader in the field of archaeology. The transformation of Rhode Island Hall as the home for the new Joukowsky Institute is a significant part of Brown's new Plan for Academic Enrichment.

The project is a leading example of the University's approach to reanimating its historic building fabric and also demonstrates its commitment to sustainability. Rhode Island Hall is the first building at Brown to be certified LEED Gold for New Construction.

Engineer: RDK Engineers; Richmond So Engineering; GZA GeoEnvironmental Inc.
General Contractor: Shawmut Design & Construction
Landscape Architect: Hines Wasser & Associates
Lighting: LAM Partners, Inc.
Owner: Brown University

工程师：RDK工程公司、里士满So工程公司、GZA土地环境公司
总承包商：肖穆特设计施工公司
景观建筑师：海因斯·瓦塞尔事务所
照明设计：LAM公司
所有人：布朗大学

Architect / 建筑师
Anmahian Winton Architects
安玛以安·温顿建筑事务所

Location / 项目地点
Providence, Rhode Island
罗德岛，普罗维登斯

Photo Credit / 图片版权
© Peter Vanderwarker
彼得·范德尔瓦尔克

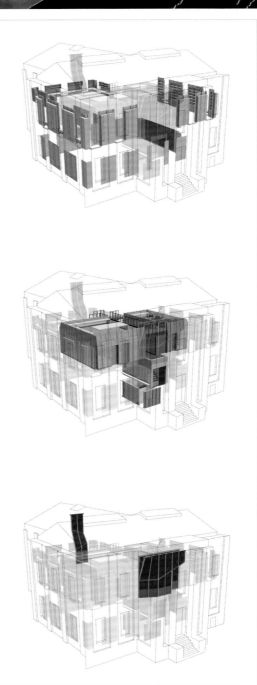

项目特色

儒科夫斯基考古和远古世纪学院重新改造了罗德岛大厅——布朗大学校园中的一座希腊复兴式建筑。作为一座捐赠的学院，它为布朗大学的研究生和本科生提供了教学、研究、实践和教室学习的空间。

项目修复了罗德岛大厅的外部，并且彻底翻新了室内。大规模的空间和结构重组让我们重新考虑了日光分布的方式。玻璃和木材的半透明感在各个功能区之间打造了不同层次的通透性和日光等级，鼓励教职员工与学生们进行互动。

历史外壳内的现代改造挑战了人们对考古学固有的印象——保守而落满了灰尘，并且让儒科夫斯基学院成为了考古学领域的领军人物。项目——将罗德岛大厅改造成儒科夫斯基学院——是布朗大学学术扩充规划的重要组成部分。

项目是大学复兴自身历史建筑网络的典范，也展示了它对可持续性设计的探索。罗德岛大厅是布朗大学第一座获得绿色建筑新工程类金奖认证的建筑。

Memory Temple

记忆殿

Jury Comments:
This project presents a remarkable approach to investigating the creation of space.
The idea of generating a form-aesthetic memory of environmental sounds by using a six-axis CNC machine that mills mapped frequencies translated into points and vectors is altogether fascinating.
The ingenious process of fabrication is visible as final product; interior space, in this case, becomes a physical manifestation of another aspect of current culture.

评委评语：
项目是对空间创造的非凡探索。
通过六轴数控机器来打磨被转换为点数和向量的映射频率来生成造型空间的想法极其吸引人。
独创的装配过程在最终产品中清晰可见；内部空间成为了当前文化的一种物理体现。

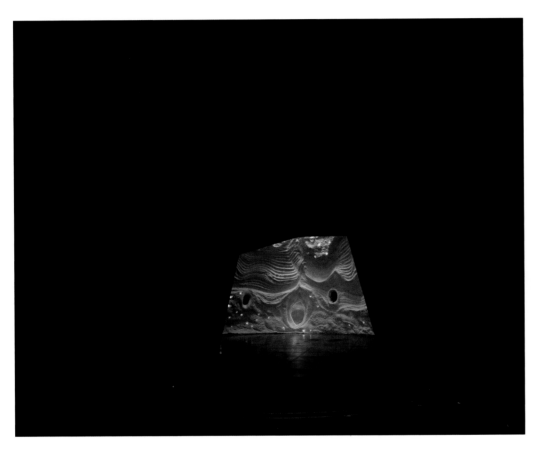
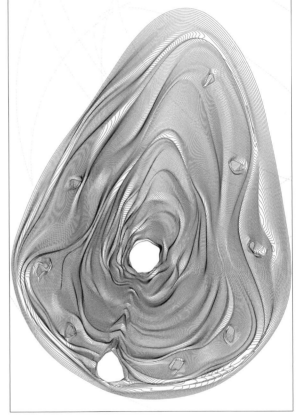

Notes of Interest

Memory Temple is an experience at the convergence of sound, material, light, form, and technology. The installation is accompanied by a site-specific composition by a world-renowned composer. The sound-scape is integral to the experience and used to explore the spatialization of sound within the physical boundaries of the gallery.

The installation proposes a new structural materiality through the use of renewable polyurethane foam. The foam was used as a total building assembly: structure, envelope, and acoustical barrier. Layers of closed cell foam (used structurally) and open cell foam (used acoustically) were combined to make up the wall assembly.

The pure geometry of the parabola provided a natural self-structural form. The musical composition is integral to the experience and provided an ever-changing mobile performed with custom software designed specifically for the installation. Resonance was exploited within the acoustically absorptive space.

Architect /建筑师
Patrick Tighe Architecture
帕特里克·泰伊建筑事务所

Location /项目地点
Los Angeles, California
加利福尼亚州,洛杉矶

Photo Credit /图片版权
© Art Gray Photography
艺术灰摄影

项目特色

记忆殿汇聚了声音、材料、光线、造型和技术体验。这一装置是世界知名设计师所打造的特定场地合成物。音景设计的运用丰富了人的体验并且在画廊的物理边界内对声音的空间化进行了探索。

装置通过可再生泡沫聚氨酯的使用提出了一种全新的结构材料选择。泡沫被应用在整个装配过程中：结构、外壳和隔音层。闭孔泡沫层（用在结构上）和开孔泡沫层（用在音效上）共同组成了墙壁整体。

抛物线纯粹的几何结构提供了天然的结构造型。乐曲成为了体验的一部分，并且通过定制的软件提供了一个不断变换的移动式演出。吸音空间内部形成了共鸣。

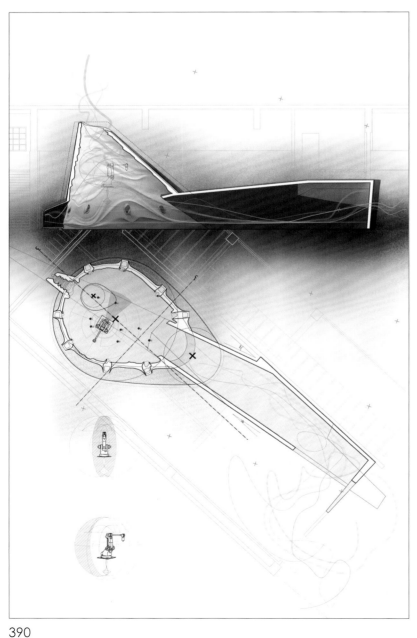

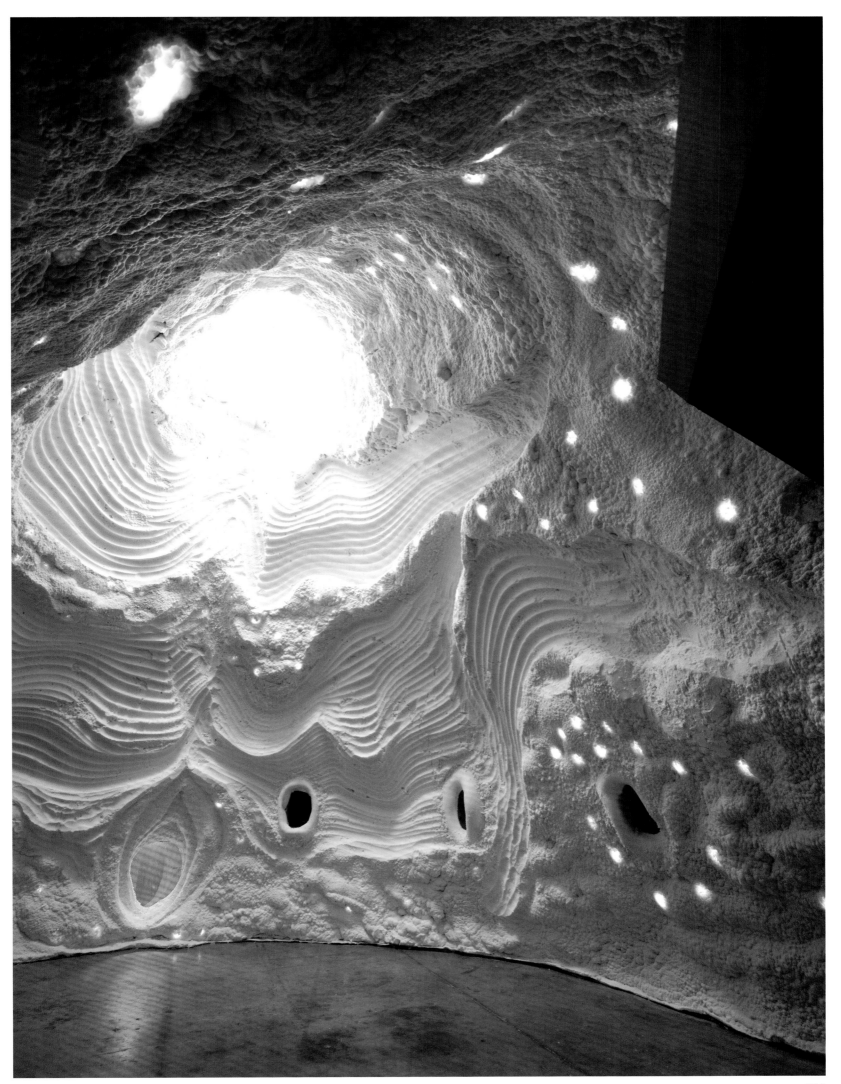

Prairie Management Group
大草原管理集团

Jury Comments:
Small and well-executed, this project is quiet, restrained, and sophisticated with a straight-forward manner towards both the composition and detail that reinforces the larger concept.
Standing in contrast to the natural prairie grass outside, the design effectively organizes the plan around a central armature inducing the glass screen walls to both modulate the plan and effectively provide light and views to the meadow beyond.

评委评语:
小巧而精致,项目安静、节制而又精妙,在整体和细节方面采用简明的设计方式来突出更大的设计理念。
设计与外面的大草原形成了对比,有效地围绕中心结构组织了布局。玻璃幕墙既调整了整体布局又引入了外面的光线和大草原的景色。

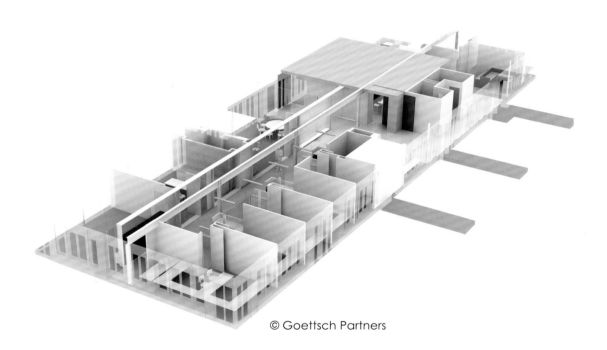

© Goettsch Partners

Notes of Interest

Executive investment offices for a retired design entrepreneur, the founder of a national home furnishings retailer, provided the opportunity to leverage the client's passion for clean, bold modern design into a dramatic, light-filled investment office environment showcasing the client's extensive glass and ceramic art collection.

Inserted into a single-story, speculative office suite, the 7,500 sf facility is organized around three compositional elements: the colonnade, created by the building's exposed structural steel columns and central ridge beam; full-height glass screen walls; and a custom maple "pavilion".

The simple, classic interior composition of thin glass frames and bold, clear millwork forms rendered in a timeless color palette – all awash in natural light – creates a platform in which the appreciation of fine art, design, and nature enables a pioneering entrepreneur to continue his lifelong passion for creating business value through design.

Engineer: Cartland Kraus Engineering, Ltd; The Structural Group
General Contractor: Pepper Construction
Landscape Architect: Hoerr Schaudt Landscape Architects
Owner: Prairie Management Group
工程师:卡尔兰德·克劳斯工程公司、结构集团
总承包商:辣椒建筑公司
景观建筑师:霍尔·肖德特景观建筑事务所
所有人:大草原管理集团

Architect / 建筑师
Goettsch Partners
杰特施事务所

Location / 项目地点
Northbrook, Illinois
伊利诺伊州，诺斯布鲁克

Photo Credit / 图片版权
© Michelle Litvin, © Goettsch Partners
米歇尔·利特文、杰特施事务所

© Michelle Litvin

项目特色

项目是由一位退休的设计行业企业家——一家美国全国家具零售商的创始人——的行政投资办公室发起。设计要求体现客户对简洁大胆的现代设计的热爱，打造一个夸张而光线充足的投资办公环境，同时也对客户的玻璃和陶瓷艺术收藏品进行展示。

近700平方米的办公空间设在一个单层的投资办公套房里，围绕着三个元素展开：由建筑的裸露钢柱和中央横梁所形成的柱廊、全高玻璃幕墙和特别定制的枫木"凉亭"。

简单而经典的室内超薄玻璃框架和大胆简洁的经典色系木工造型沐浴在阳光之中，形成了欣赏艺术品、设计和自然的平台，让这位先锋企业家能够继续自己的终身爱好，通过设计创造商业价值。

Record House Revisited
实录房重建

Jury Comments:
An excellent example of new work within a significant mid-century modern structure, the interventions appear to reinforce the original design concept.
Eliminating carefully selected interior walls allows floor-to-ceiling openings, emphasizing the integrity of the two pavilions.
The new work serves to highlight the naturally lit passage and accentuate the overall spirit of the house.

评委评语:
作为对20世纪中期现代住宅的杰出改造,项目凸显了原始设计的概念。
拆除精挑细选的室内墙壁为落地窗提供了空间,强调了住宅两个部分之间的完整性。
翻新工作突出了自然光照的走廊并着重描绘了住宅的整体精神。

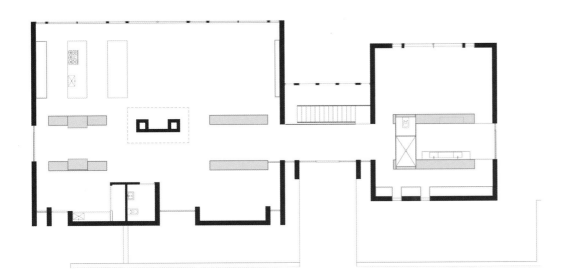

Notes of Interest
Four decades after their project was featured in the 1969 Record Houses issue of Architectural Record, the owners sold the house to a young couple. A condition of the sale was that the new owners would respect the character of the project, yet be able to revisit and alter the contained quality of the interior rooms to create a continuous living space visually connected to the woodland site.
An analysis of the existing structure revealed ordering devices through which the new work could be understood. A truss roof system allowed interior walls to be eradicated, yielding a condition of an unencumbered public and private pavilion linked together by a glass entry node. Floor to ceiling window apertures relating the pavilions could not be experienced within the original floor plan.

General Contractor: Prr
总承包商: 雷集团

Architect / 建筑师
David Jameson Architect
大卫·詹姆森建筑事务所

Location / 项目地点
Owings Mill, Maryland
马里兰州，奥因斯米尔

Photo Credit / 图片版权
© Paul Warchol
保罗·瓦尔孔

Registering the new work to the existing house is a conceptual allee of walnut casework. The casework weaves together and provides clarity to the various living areas. The quarter sawn casework and flat sawn flooring employ walnut in a Chiascuro manner, creating bold contrasts to the existing white painted brick walls and plaster ceiling. Corian casework elements are positioned as kitchen, mudroom, and bath objects, further juxtaposing a smoothness to the textural brick and plaster. The purity of the original brick fireplace and skylight ring at the center of the house is exposed and left uninterrupted, allowing for additional connection to the site.

项目特色

在被《建筑实录》选为1969年的实录房之后的40年,屋主将住宅卖给了一对年轻夫妇。房屋的新主人希望尊重项目的原有特色,同时也调整室内空间的布局来打造连续统一的起居空间,使其与周边的林地在视觉上联系起来。

对原有结构的分析显示:通过新工程来排列空间是可行的。桁架式屋顶结构让房屋摆脱了室内墙壁,打造了无障碍的公共和私人空间,使二者通过玻璃入口隔开。起到连接作用的落地窗是原始平面规划中所没有的。

新旧空间以胡桃木柜子连接起来。它们交织在一起,明确了各个起居区之间的界限。纵切的柜子与横切的地板采用了不同的方式来使用胡桃木,与原有的白漆砖墙和石膏天花板形成了鲜明对比。可丽耐柜子元素被设置成厨房、前厅、浴室等空间,进一步与砖块和石膏形成了对比。

原始砖块壁炉的纯净感和房间中央的天窗被完整地保留下来,促进了项目与场地的额外联系。

The Wright at the Guggenheim Museum
古根海姆博物馆莱特餐厅

Jury Comments:
This project is sensitively handled and respectful of the essence of the original architecture.
With the confined space and ostensibly modest budget, given those challenging constraints, this project is exceptional. Of special note is the programmatic flexibility.
The design approach was controlled but playful, and complements the nuance of the museum overall movement and dynamic.

评委评语:
项目巧妙地应对并尊重了原始建筑结构。
在有限的空间和预算之中,项目达到了非凡的效果。最明显的就是项目的灵活性。
设计方法谨慎而巧妙,为博物馆带来了精妙的动态之感。

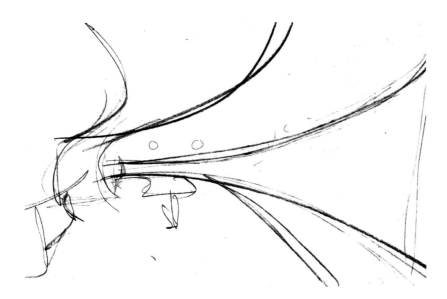

Notes of Interest
It was both an incredible honor and an exhilarating challenge to create The Wright, the new restaurant at the Guggenheim Museum – which is the first addition to the building's iconic interior. The architect sought to create a contemporary response to complement the building with an extremely modest budget and 1,600 square feet in which to work.

The design solution references the building's architecture, what Wright specifically called "the primitive initial", without repeating it. In the process, the architects transform underlying architectural geometries into dynamic spatial effects. The sculptural forms create a flared ceiling. The undulating walls become comfortable seating. The arced bar and communal table animate the space. The playfulness of these forms offers a dynamic experience for visitors.

This project is highly tactile and crafted from innovative, contemporary materials. These include fiber-optic layered walnut, a shimmering skin of innovative custom metalwork, seamless Corian surfaces, illuminated planes of woven grey texture, and a glowing white canopy of layered taut membranes. Together these materials and colors form a perfect complement to the site-specific artwork by Liam Gillick. The surfaces and textures embody movement, creating an ever-changing aesthetic that is enlivened with subtle layers of illumination and glowing tiers of light that envelope the room.

The space achieves an elegant and dynamic setting for dining that both celebrates the museum and transcends it.

Engineer: HHF Design Consulting, Ltd.
General Contractor: James G. Kennedy & Co., Inc.
Lighting: Tillotson Design Associates
Owner: Restaurant Associates
工程师: HHF设计咨询公司
总承包商: 詹姆斯·G·肯尼迪公司
照明设计: 迪洛森设计事务所
所有人: 餐厅联合公司

Architect / 建筑师
Andre Kikoski Architect, PLLC
安德里·吉果斯基建筑事务所

Location / 项目地点
New York City, New York
纽约州，纽约

Photo Credit / 图片版权
© Peter Aaron
彼得·亚伦

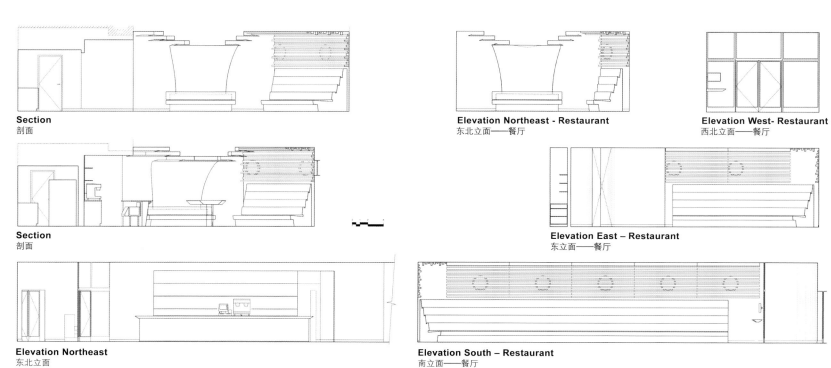

Section
剖面

Section
剖面

Elevation Northeast
东北立面

Elevation Northeast - Restaurant
东北立面——餐厅

Elevation East – Restaurant
东立面——餐厅

Elevation South – Restaurant
南立面——餐厅

Elevation West- Restaurant
西北立面——餐厅

项目特色

为古根海姆博物馆来设计莱特餐厅是一项不可思议而令人欣喜的挑战,这是博物馆内部的第一个新加设施。建筑师力求以极少的预算在这个150平方米的空间内打造一家现代的餐厅。

设计参考了建筑的结构,却又不重复表现。在设计过程中,建筑师将潜在的建筑几何造型改造成了动感的空间效果。雕塑造型打造了向外展开的天花板。波浪起伏的墙壁成为了舒适的座椅。弧形吧和公共桌让空间充满了生气。这些造型的巧妙运用为访客提供了动态的体验。

项目具有高超的触感,采用创新型现代材料建成。其中包括光纤层胡桃木、闪闪发光的创新定制金属制品、无缝的可丽耐平面、编织的灰色发光板和发光的彩色叠层天花板。这些材料和色彩形成了一个完美的博物馆补充结构。装饰表面和纹理体现了运动感,营造了一个千变万化的美感。灯具和发光体的微妙层叠让光线洒满了整个房间。整个空间打造了优雅而充满活力的就餐环境,既赞美了博物馆又超越了它。

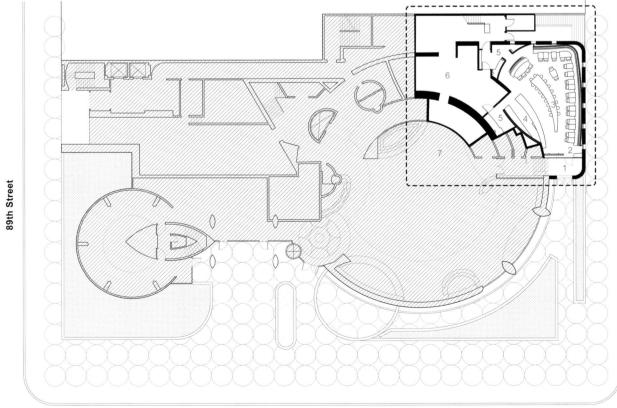

Site Plan:	总平面图:
1. Vestibule	1. 门廊
2. Hostess	2. 女服务员休息室
3. Dining Room	3. 餐厅
4. Bar	4. 酒吧
5. Service	5. 服务区
6. Kitchen	6. 厨房
7. Not In Scope	7. 不在项目范围内

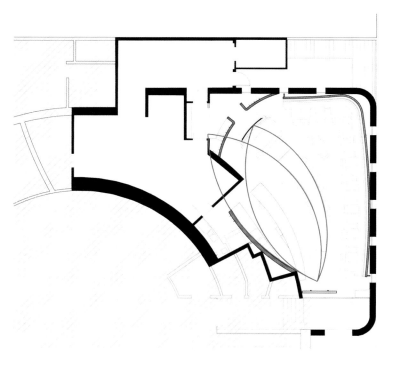
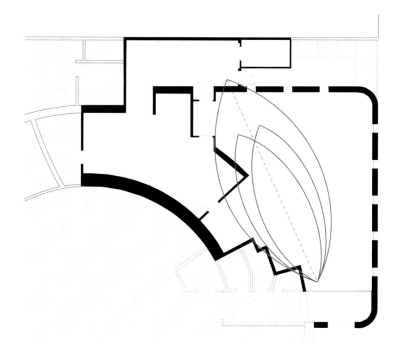

Fayetteville 2030: Transit City Scenario

费耶特维尔2030：城市交通情景规划

Jury Comments:
The premise of this project is very forward-thinking.
There is a great appreciation for the comprehensive scope of this project as well as the clear visualization of the character of the public space within the scale of infrastructure. The preservation of the rural character of the existing town as well as the addition of the more modern elements has been masterfully handled.

评委评语：
项目的前提在于高瞻远瞩。
项目的整体规划和清晰的公共空间规划大受赞赏。
原有城区的乡村特色与新建区域所具有的现代元素被巧妙地结合在一起。

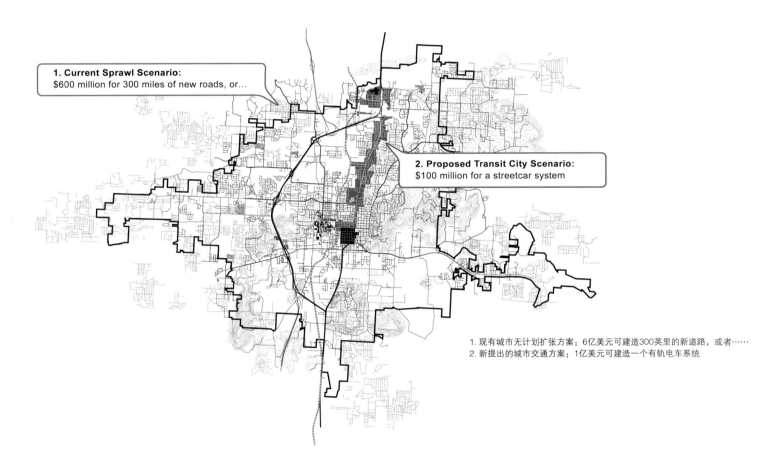

1. 现有城市无计划扩张方案：6亿美元可建造300英里的新道路，或者……
2. 新提出的城市交通方案：1亿美元可建造一个有轨电车系统

Notes of Interest

As a complement to Fayetteville's 2030 City Plan, a 2030 Transit City Scenario Plan independently models a future based on development of a streetcar system. While city planning is generally future-oriented, scenario planning models specific futures from the insistent exploration of a particular driver through "what if" propositions.

Scenario planning helps the community envision plausible planning possibilities that would not have emerged from charrettes and similar participation processes. They ask: what if 80% of future growth was incented to locate around a street car system proposed for Fayetteville's main urban arterial? The city could create a five-mile signature multi-modal transit boulevard, transforming underperforming development into mixed-use transit-oriented neighborhoods.

The post-carbon sustainable city will be based on multi-modal transportation systems that support passenger rail and walkable neighborhoods in tandem with the automobile, and their objective was to prepare such a model for a five-mile segment between the downtown/university district and the regional shopping mall at the city's edge.

Akin to the early 20th century streetcar cities, built at 7-14 dwelling units per acre, their plan updates the role of rail transit in reforming land uses for small town markets.

Owner: City of Fayetteville, Arkansas
所有人： 阿肯色州费耶特维尔市

Architect / 建筑师
University of Arkansas Community Design Center
阿肯色大学社区设计中心

Location / 项目地点
Fayetteville, Arkansas
阿肯色州，费耶特维尔

Photo Credit / 图片版权
© University of Arkansas Community Design Center
阿肯色大学社区设计中心

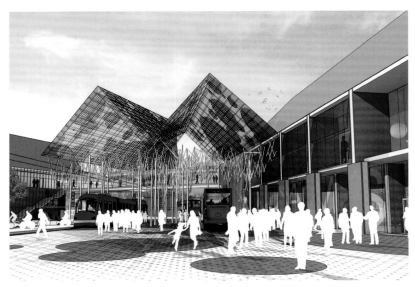

项目特色

作为费耶特维尔2030年城市规划的补充部分，2030年城市交通方案以有轨电车系统为基础，建立了独立的设计模型。城市规划普遍面向未来，而情景规划模型则坚持不懈地通过"假设"前提来探索特定的驱动方案。

情景规划帮助社区预想合理的规划可能性，与专家研讨会和类似的流程不同。我们提出：假设80%的未来开发项目都围绕有轨电车系统展开会怎样？城市将打造一个8公里长的多模式交通大道，将表现不佳的开发工程变成多功能的便利社区。

后碳时代可持续城市将以多模式交通系统为基础。客运铁路和步行街区将与机动车道相结合。我们的目标是在市中心/大学区域与城郊的区域购物中心之间建设一条长达8公里的路段。

20世纪初拥有有轨电车的城市在每英亩（约4,047平方米）空间内建造7-14套公寓，与它们类似，我们的规划将突出有轨电车在改良土地使用方面的作用。

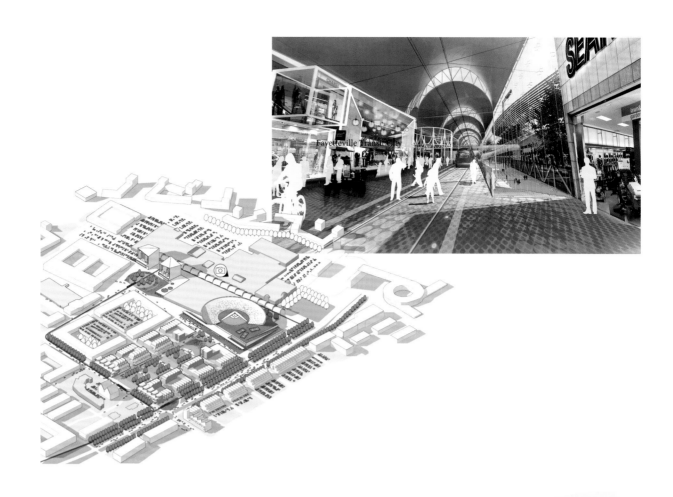

Grangegorman Urban Quarter Master Plan
格兰格曼城区总体规划

Jury Comments:
This project presents an impressively comprehensive approach to site planning, with clear and systematic design strategies. The adaptive reuse of the historic existing buildings appears to be very successful.
The sense of transparency both within the new buildings and through the interstitial public space is impressive. The diagrammatic representation of the fundamental design strategies was both clear and compelling.

评委评语：
项目呈现了非凡的综合场地规划，拥有明确而系统的设计策略。对原有历史建筑的修葺改造看起来相当成功。
新建筑以及间隙公共空间的通透感十分出色。基本设计策略的图解展示清晰而有力。

Notes of Interest

The Grangegorman Master Plan represents the largest higher-education campus development ever undertaken in the history of the state of Ireland, creating a vibrant new Urban Quarter for Dublin's north inner city.

It will accommodate 422,300 square meters of academic and residential buildings for the Dublin Institute of Technology (DIT), along with replacement psychiatric facilities and new primary care facilities for Ireland's national health care service, the HSE, and new amenities for the local community and the wider surrounding city.

The site is 73 acres which is currently used by the old St. Brendan's Psychiatric Hospital. It has been walled off from the rest of the city of Dublin since the early nineteenth century and is one of the largest undeveloped pieces of land in the city.

Engineer: ARUP
Environmental Sustainability: Battle McCarthy Consulting Engineers
Landscape Architect: Lützow 7
Conservation: Shaffrey Associates
Owner: Dublin Institute of Technology and Health Service Executive

工程师：ARUP
环境可持续设计：贝托·麦克卡尔西工程咨询公司
景观建筑师：卢佐7
历史保护：沙弗里事务所
所有人：都柏林理工大学和医疗服务执行委员会

Architect / 建筑师
Moore Ruble Yudell Architects & Planners; DMOD Architects
摩尔·卢堡·尤戴尔建筑规划公司；DMOD建筑事务所

Location / 项目地点
Dublin, Ireland
爱尔兰，都柏林

Photo Credit / 图片版权
© Moore Ruble Yudell Architects & Planners
摩尔·卢堡·尤戴尔建筑规划公司

项目特色

格兰格曼总体规划呈现了爱尔兰历史上最大规模的校园开发项目，它为都柏林的北部内城打造了一个全新的城区。

项目将为都柏林理工大学提供422,300平方米的教学和住宿空间，并且重新为爱尔兰国家医疗服务执行委员会设置了精神院和全新的初级护理设施，同时还为当地社区和更广泛的城市区域提供了全新的便利设施。

整个项目场地近30公顷，原来被圣布兰丹的精神病院所占据。在19世纪早期开始，它就与都柏林市的其他区域隔开，是市内最大的一片未开发区域。

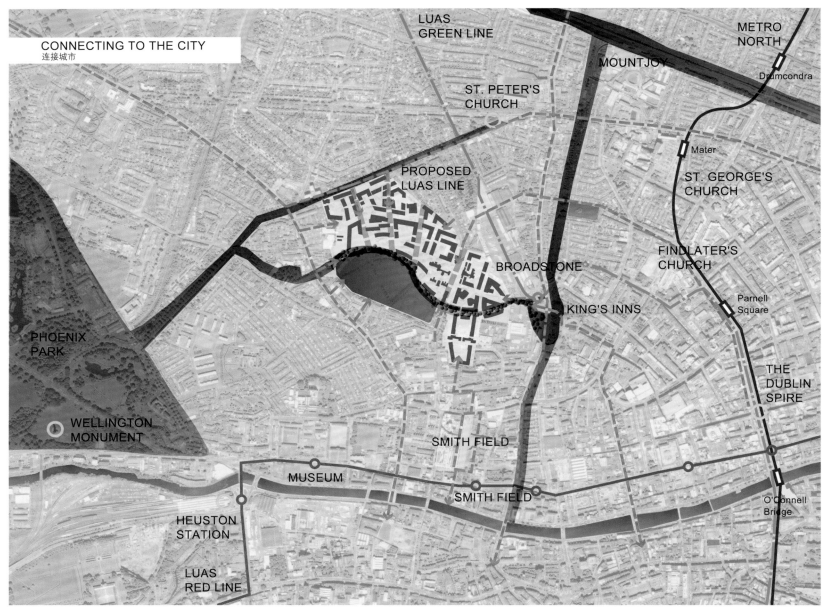

CONNECTING TO THE CITY
连接城市

Summer Sun Path
夏季太阳轨道

Winter Sun Path
冬季太阳轨道

South-West Prevailing Winds
Wind Harvesting/Evaporative Cooling
西南盛行风
风力收集/蒸发冷却

Landscape Buffer - Protection From Cold Winter Winds
景观缓冲区——保护冬季寒风

Stormwater Management Water Retention Pond
雨水处理
保水池塘

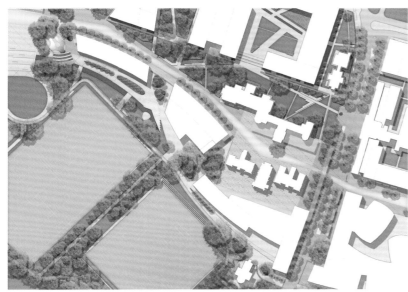

Jordan Dead Sea Development Zone Master Plan
约旦死海开发区总体规划

Jury Comments:
The jury appreciated the project's emphasis on the creation of a public waterfront, unusual in this region of resorts.
The development of a planning strategy that is structured around the movement of water demonstrates an ecological sensitivity.
The master plan skillfully accommodates different populations and is incredibly comprehensive.

评委评语：
评委欣赏项目对打造公共水畔的重视，在同类地区中很少见。
规划策略的开发围绕着水的运动展开，具有生态敏感性。
总体规划巧妙地结合了不同人群，极具综合性。

Notes of Interest

In order to capitalize on its cultural and political assets, the Government of Jordan has established a series of six Development Zones in which directed efforts are being made to increase foreign and domestic investment. The Dead Sea Development Zone encompasses 40 square kilometers of coastal land along the lowest body of water on earth.

The Detailed Master Plan for the Dead Sea Development Zone lays out a vision and blueprint for fostering a dynamic, robust, and sustainable tourism-based economy at the Dead Sea that will become a source of pride and revenue for the Kingdom and set the highest standard for sustainable development and innovative urban design. Critically, the plan establishes a "balanced approach" between development and conservation of this most precious resource. At the same time, it will strengthen local economies and greatly support social infrastructure for nearby existing communities. Comprehensive design guidelines, a detailed infrastructure report, and an extensive market study enable the client to attract investment. Future development will follow a carefully choreographed phasing plan that capitalizes on existing investments, introduces infrastructure to precede development, and preserves large contiguous tracts of developable land as future land banks.

The year-long master planning process focused on critical work sessions held in Amman with local community leaders, government officials, international and local investors, and leading academics.

Collaboration: Buro Happold, Sigma Consulting, Tetra Tech
Owner: Jordan Development Zones Commission
合作： 布罗·哈波尔德；西格玛咨询公司；德照科技
所有人： 约旦开发区委员会

项目特色

为了充分利用文化和政治资产，约旦政府建立了六个开发区，以吸引国外和本国的投资。死海开发区环绕着40平方公里的全球最低水体的沿海区域。
死海开发区的详细城市规划为打造死海区域活跃、强健而可持续的以旅游业为基础的经济描绘了蓝图。不久的将来，死海将成为约旦为之自豪的经典，为可持续开发和创新城市规划建立最高的标准。该规划在开发和资源保护之间建立了"均衡的处理方式"。同时，它将会促进当地经济发展，大力支持附近区域的社会基础建设。
综合设计方针———一份详细的基础设施报告和一个广泛的市场研究保证了客户能够吸引投资。未来的开发将遵循精心设计的阶段性规划，利用现有投资、在前期开发中引入基础设施并且保护大片的可开发土地。
历经一年的总体规划进程聚焦于在安曼举行的工作研讨会。研讨会将聚集当地社区领袖、政府官员、国际和本地投资商以及顶尖学者。

Architect	Location	Photo Credit
Sasaki Associates, Inc.	Amman, Jordan	© Sasaki Associates, Inc.

OPEN SPACE STRATEGY

Natural assets from the framework for the Master Plan's open space strategy. Rivers, wadis, steep cliffs, and sensitive habitats are removed from the inventory of developable land.
Landmark public open spaces are planned throughout the project site, providing public access to the sea at every opportunity: a large public park, eco-tourism zones, pedestrian paths, and urban plazas.

开放空间策略

总体规划的开放空间策略中的自然资产。河流、干河床、悬崖和敏感的栖息地被从可开发的土地上移走。地标性公共开放空间遍布整个项目场地,提供了许多入海通道:大型公园、生态旅游区、人行道和城市广场。

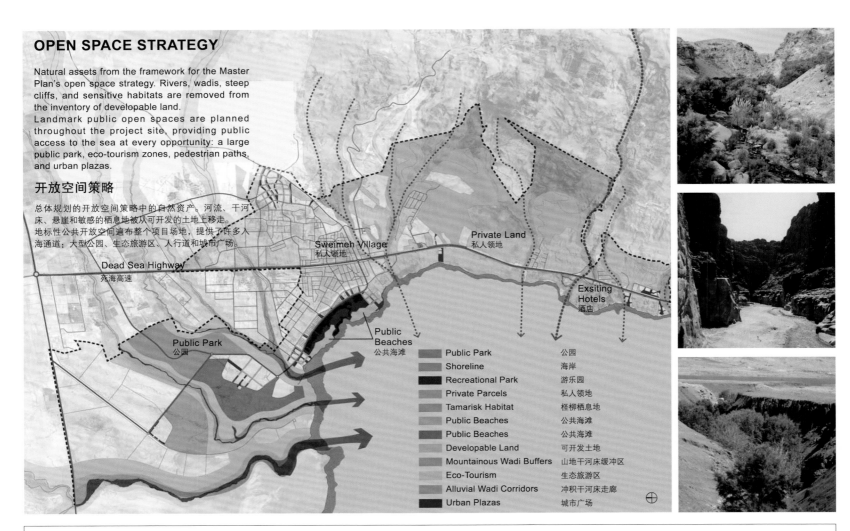

UNDERSTANDING NATURAL HYDROLOGY

The Jordan River and several smaller alluvial rivers terminate at the northern shore of the Dead Sea. Along the eastern banks of the Sea, the landscape transitions into rocky cliffs, through which mountainous wadis cut deep, dramatic incisions. It is critical to protect these complex ecological systems.

理解自然水文地理

约旦河和死海北岸一些较小的冲积河都在死海北岸中止。而死海东岸的景观变成了岩石悬崖,山地干河床通过悬崖切开了深深的切口。对这些复杂的生态系统的保护至关重要。

冲积干河床:
1. 干河床和河岸的 100 米范围之内不允许进行任何开发。
2. 河湾外半径陡峭、不稳定的河岸易于坍塌。
3. 河湾内半径的缓坡河岸。
4. 随着死海的水平面下降而深化切口。

山地干河床:
5. 山脊和山脊之间不允许进行开发,为了保护分水岭山脊不触碰干河谷水域的界限。
6. 干河床口冲积扇。
7. 山脊与山脊之间的缓冲区。

ALLUVIAL WADIS
No development permitted within 100 meters of wadi and river banks
1 — 100 m
2 — Steep, unstable banks along the outer radii of river bends are prone to collapse.
3 — Shallowly sloped banks along the inner radii of river bends
4 — Deepening incisions as the Dead Sea level drops

MOUNTAINOUS WADIS
No development permitted from ridgeline to ridgeline in order to protect watersheds
7 — Ridgeline to ridgeline buffer zone
5 — Ridge-Lines form the boundary of the wadi's watershed.
6 — Alluvial fan at Wadi Mouth

- Wadis with Regional Watersheds — 干河床和地区水域
- Wadis with Local Watersheds — 干河床和当地水域
- Alluvial Wadis — 冲积干河床
- Project Site Boundary — 项目场地边界

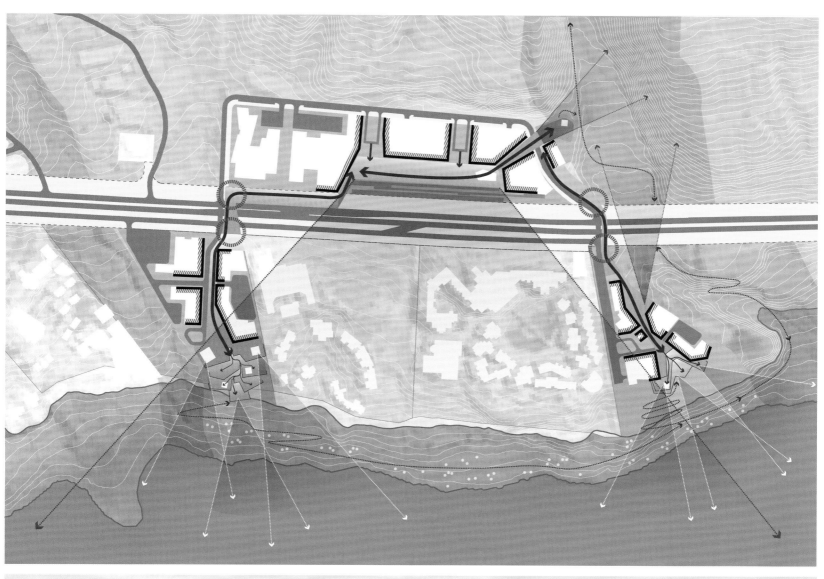

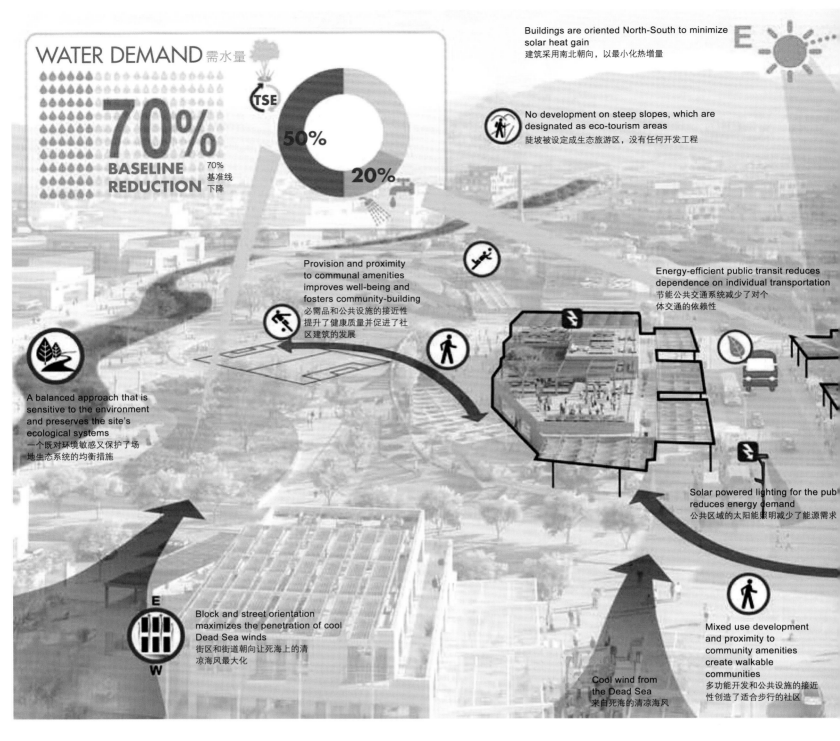

EXISTING WATER CYCLE
现有的水循环

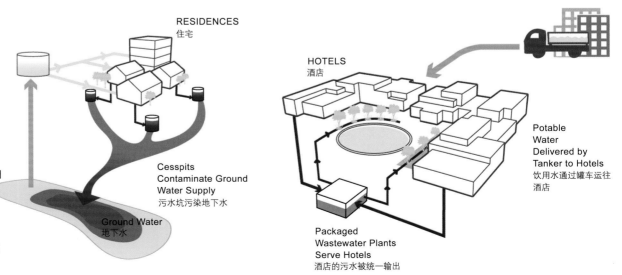

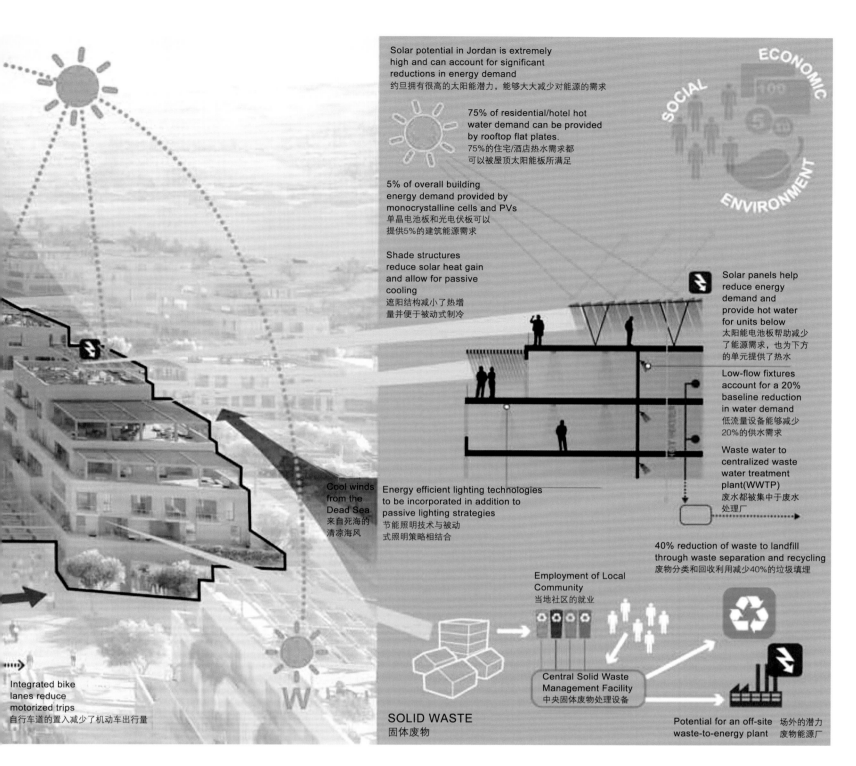

PROPOSED WATER CYCLE
计划的水循环

Potable Water Demand is lessened and Water Quality is improved
饮用水需求减少，而水的质量得到了提高

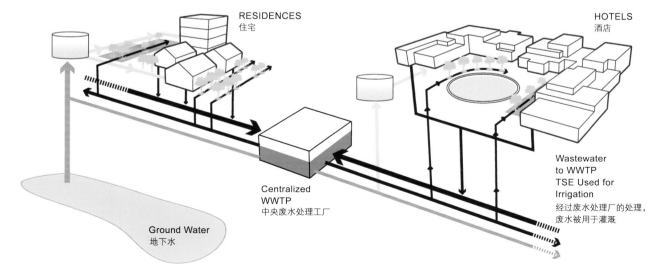

Master Plan for the Central Delaware
德拉瓦河中段总体规划

Jury Comments:
This is a very bold, long term vision.
The project demonstrates great connectivity back into the neighborhood fabric, integrating both existing buildings, developed open spaces, and the esplanade walk.
A good range of density has been represented, particularly along the river, and this plan has transformed the city in a substantial way.

评委评语：
这是一个非常大胆而长期的规划。
项目展示了街区网络的连通性，与原有建筑、已开发的开放空间和散步走道结合在一起。
项目呈现了良好的密度结构，特别是沿河区域，这一规划以充实的方式改变了城市。

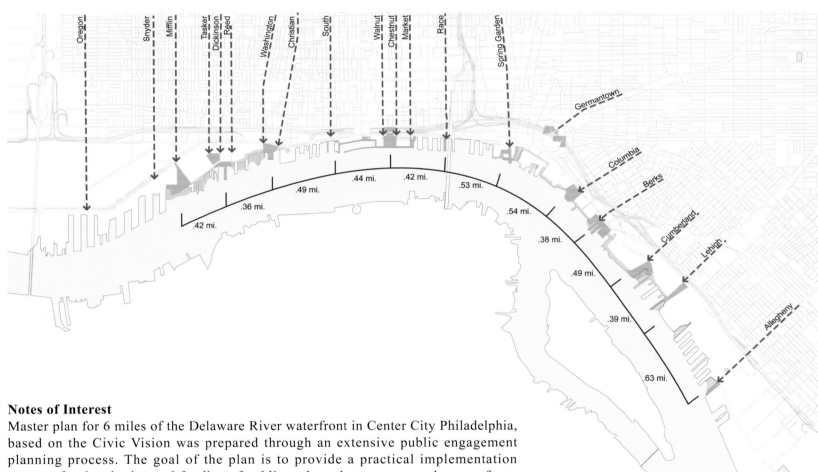

Notes of Interest
Master plan for 6 miles of the Delaware River waterfront in Center City Philadelphia, based on the Civic Vision was prepared through an extensive public engagement planning process. The goal of the plan is to provide a practical implementation strategy for the phasing and funding of public realm enhancements to the waterfront, including the locations of parks, a variety of waterfront trails, and connections to existing upland neighborhoods.

Specific zoning recommendations to shape private development as well as design guidelines for the public spaces are integral components of this project. A crucial aspect of the planning process was the creation of a dialogue with stakeholders and the incorporation of public input.

Throughout the project, the team had direct contact with neighborhood associations, elected officials, heads of major state and local agencies, as well as property owners and developers. At the end of each phase, the team conducted a public meeting to solicit feedback and engage constituents in the work's progress.

项目特色
费城德拉瓦河畔绵延近千米的总体规划项目以市民为基础，公众广泛参与了规划过程。目标是为河畔区域提供一个具有实用价值的阶段规划策略，包括公园的选址、各种河畔走道设计及与原有山地居民区进行连接。
项目还包括私人开发工程的特别分区建议和公共区域的指导设计方针。规划流程的决定性部分在于打造利益相关者和公共意见之间的互动。
设计团队在项目过程中与周边居民团体、官员、当地的公司巨头以及房产所有人和开发商进行了直接的交流。在每个阶段的终点，设计团队都组织了一次公众见面会来征求反馈意见，进一步保证工作的流程。

Associate Architect: Kelly/Maiello Architects & Planners
Consultant: CHPlanning; Greater Philadelphia Urban Affairs Coalition; HR&A Advisors; Hurley Franks & Associates
Cost Estimator: Davis Langdon
Engineer: KS Engineers, P.C.; Parsons Brinckerhoff
Land Use Counsel: Blank Rome LLP
Programming: Karin Bacon Enterprises
Waterfront Policy Advisor: Toni L. Griffin
Owner: Delaware River Waterfront Corporation
合作建筑师： 凯利/梅罗建筑规划事务所
顾问： CH规划、大费城城市事务联盟、HR&A顾问公司、赫利·弗兰克斯事务所
成本估算： 戴维斯·朗登
工程师： KS工程公司；柏诚集团
土地使用顾问： 博锐律师事务所
规划： 卡琳·培根公司
滨水政策顾问： 托尼·L·格里芬
所有人： 德拉瓦河畔开发公司

Architect /建筑师
Cooper, Robertson & Partners, KieranTimberlake, OLIN
库伯·罗伯森事务所、基兰·廷伯莱克、OLIN

Location /项目地点
Philadelphia, Pennsylvania
宾夕法尼亚州，费城

Photo Credit /图片版权
© Brooklyn Digital Foundry
布鲁克林数码制造厂

Miami Beach City Center Redevelopment Project
迈阿密海滩城市中心再开发项目

Jury Comments:
Compelled with this small project with big impact, the building and its public park space has developed an extremely vital and dynamic cultural space within the city.
The inventive use of cultural programming by the symphony hall; the outdoor concert projections in the park, have made culture accessible to everyone.
The project makes good use of existing buildings and creates a strong connection back to the city, energizing and activating Miami Beach.

评委评语：
小项目拥有大影响，建筑和公园空间共同在城市中打造了一个充满活力的文化空间。
项目创造性地利用了交响乐厅来举办文化项目；公园里的露天音乐会投影让文化走向了基层。
项目充分利用了原有建筑，并创造了与城市的紧密联系，活跃了整个迈阿密海滩地区。

Consultant: Acoustic Dimensions; Prosound and Video; Sonitus, LLC; Theatre Projects Consultants
Engineer: Coastal Systems International; Cosentini Associates; Douglas Wood Associates; Gilsanz, Murray, Steficek, LLP; Kimley Horn and Associates, Inc
General Contractor: Facchina Construction of Florida, LLC
Landscape Architect: Raymond Jungles Associates; Rosenberg Gardner Design
Lighting: LAM Partners, Inc.
Seating: Poltrona Frau
Acoustical: Nagata Acoustics America, Inc
Owner: City of Miami Beach; New World Symphony

顾问： 声学维度公司；宝笙视听；索尼托斯公司；剧院项目顾问
工程师： 沿海系统国际公司；科森蒂尼事务所；道格莱斯木业；GMS公司；吉姆利·霍恩事务所
总承包商： 佛罗里达法希纳建筑公司
景观建筑师： 雷蒙德·金格斯事务所；罗森博格·加德纳设计
灯光设计： LAM事务所
座椅设计： 珀尔特罗那·弗劳
音响设计： 永田音响美国公司
所有人： 迈阿密市政府；新世界交响乐团

Notes of Interest

Opened in January 2011, the 5.86-acre Miami Beach City Center Redevelopment project consits of New World Center, an innovative facility for music education and performance with state-of-the-art technical capabilities; Miami Beach SoundScape, an adjacent 2.5-acre public park and event space; and a 556-space municipal parking structure.

The project is located on two city blocks previously used as surface parking lots. New World Center is a unique performance, education, production, and creative space with state-of-the-art capabilities, owned and operated by the New World Symphony (NWS). A global hub for creative expression and collaboration, and a laboratory for the ways music is taught, presented and experienced, the building enables NWS to continue its role as a leader in integrating technology with music education and concert presentation. It is used by NWS for educational, concert, and broadcast activities. The building features a giant, 7,000-square-foot projection wall used for outdoor presentations to audiences in the adjacent park, complemented by an immersive audio system in the outdoor viewing area.

Miami Beach SoundScape is a multi-use park that serves as an urban oasis and a gathering place for cultural and special events. It is a unified expression of passive recreation, pleasure and culture – a space that supports a multitude of day and night uses that, combined with New World Center's expansive projection wall, marries music, design, landscape and community.

项目特色

占地2.37公顷的迈阿密海滩城市中心再开发项目于2011年1月开启，其中包括新世界中心（创新型音乐教育和演出设施，配有先进的技术设施）、迈阿密海滩音景（近1公顷的公园和活动空间）和拥有556个车位的市政停车场。

项目所在的两个街区从前是地上停车场。新世界中心是一个独特的表演、教育、制作和创意空间，配有最先进的技术设施，由新世界交响乐团运营。作为一个创意表达和合作中心以及一个音乐教育的实验室，建筑让新世界交响乐团延续了自己作为音乐教育和音乐会制作的领军人物。它被新世界交响乐团用作进行教学和举办音乐会和广播活动的中心。650平方米的投影墙向外面公园中的观众进行展示，并搭配一个沉浸式露天视听系统。

迈阿密海滩音景是一个多功能公园，是一片绿洲，能够作为文化活动和特殊活动的集会场所。它结合了被动式休闲、娱乐和文化，支持大量的日间和夜晚活动。公园与新世界中心的投影墙一起，集音乐、设计、景观和社区活动于一身。

Architect / 建筑师	Location / 项目地点	Photo Credit / 图片版权
Gehry Partners, LLP; West 8 Urban Design and Landscape Architecture 盖里事务所；西8城市规划和景观建筑事务所	Miami Beach, Florida 弗罗里达州，迈阿密海滩	© Gehry Partners, LLP, © West 8 New York, © Emilio Collavino, © WorldRedEye, © Claudia Uribe, © Robin Hill, © Craig Hall 盖里事务所、西8纽约公司、艾米里奥·克拉维诺、克劳迪娅·乌里贝、罗宾·希尔、克雷格·霍尔

Portland Mall Revitalization
波特兰林荫大道复兴工程

Jury Comments:
Beautiful! This project asserts that urban design can really work, and exemplified this through design at both the large scale and the detail.
A strong sense of urban space has been created by the continuity of the streetscape. The environment is accessible on all levels.
It has been beautifully executed, with fine design details upgrading the good bones of the existing situation. A sense of urban play and connectivity is evident.

评委评语：
漂亮！项目说明了城市规划的现实价值，通过大规模设计和细部设计展示了自己。
项目通过街景的延续性凸显了强烈的城市空间感。
设计的执行十分完美，精致的细节设计提升了原有框架的等级，城市感和连通性都异常明显。

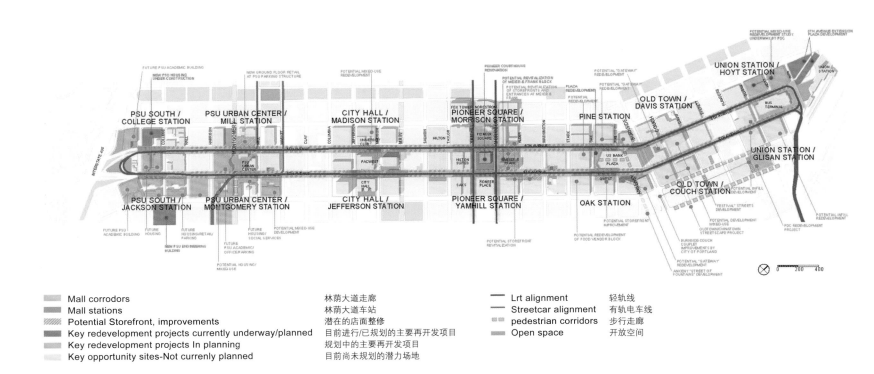

Notes of Interest

Already an icon of progressive planning, the revitalized Portland Mall is significant nationally as a new benchmark in urban design, place-making and transportation infrastructure. Extending the entire length of downtown Portland, it mixes multiple modes of transportation, stimulates adjacent development and re-establishes itself as one of Portland's premier civic spaces.

The project involved renovation or rebuilding of 58 blocks and 59 intersections while providing exclusive transit lanes for bus and light rail, dedicated lanes for autos and bicycles, enhanced sidewalks for pedestrians, and parking and loading zones.

The original Mall created in 1978 was deteriorating after years of heavy use and diminishing public maintenance funding. To reverse this decline and accommodate a projected 1 million new residents by 2030, TriMet in partnership with the City of Portland, Metro, the Portland Business Alliance and the architect, teamed up to create a vision for the Mall as a Great Street.

After more than a decade of process, design and construction, the Mall's reopening was complemented by two major hotel redevelopments, national retail openings, over 40 local storefront renovations, and several institutional projects.

The revitalized Mall combines design character, aspirations, active context, operations and management of a truly great street for the 21st century.

Project Manager: Shiels, Obletz, Johnsen
Engineer: Dewhurst MacFarlane and Partners; John Knapton; KPFF Consulting Engineers, Inc.; LTK Engineering Services; URS Corporation
General Contractor: Kiewit Construction Group, Inc.; Stacy and Witbeck, Inc.
Landscape Architect: Mayer/Reed
Owner: Tri-County Metropolitan Transportation District of Oregon (TriMet)

项目经理： 希尔斯；奥博勒兹；约翰逊
工程师： 德赫斯特·马克法尔莱恩事务所；约翰·耐普顿；KPFF工程咨询公司；LTK工程服务；URS公司
总承包商： 基威特建筑集团公司；斯塔希和惠特拜克公司
景观建筑师： 迈尔/里德
所有人： 俄勒冈三郡都市交通区

Architect / 建筑师
ZGF Architects LLP
ZGF建筑事务所

Location / 项目地点
Portland, Oregon
俄勒冈州，波特兰

Photo Credit / 图片版权
© Bruce Forster
布鲁斯·福斯特

项目特色

作为进步规划的标杆，复兴的波特兰林荫大道成为了美国城市规划、场所营造以及交通基础设施的新标准。林荫大道穿越了波特兰市中心，混合了多种交通模式，刺激了沿线的开发并且重新确立了自己在波特兰市政空间中的领军地位。

项目涉及58个街区和59个交叉路口的翻新或重建，同时为公交和轻轨提供了专属车道，划分了机动车道和非机动车道，改造了人行道、停车场和装载区。

林荫大道最初建于1978年，在多年的使用和疏于维护之后逐渐破败。为了逆转它的衰退并在2030年前为100万居民提供住宅，三郡都市交通区与波特兰市政府、地铁、波特兰商业联盟以及建筑师共同为林荫大道打造了全新的规划。

在超过十年的处理、设计和建设过程之后，林荫大道重新开放，其沿线配有两座主要酒店的重建、国家级零售店的开张、40多个街面翻新以及一些公共项目的建造。

复兴后的林荫大道集设计特色、启发性和活跃的环境于一身，是21世纪街道管理和运营的典范。

Reinventing the Crescent: Riverfront Development Plan

改造新月区：河畔开发规划

Jury Comments:
This is an innovative and radical approach to readdressing the levee on the Mississippi and reconnecting the citizens of New Orleans back to their riverfront.
The typologies that are being developed will transform the visual and physical connection of the city to the river.
The use of existing programmatic institutions and amenities to focus development along the river is particularly laudable.

评委评语：
这是一个创新型辐射设计，重新定位了密西西比河的大堤，将新奥尔良的市民与河畔重新联系了起来。
项目将改变城市与河流之间的视觉与地理连接。
项目将原有功能建筑和便利设施与河畔开发结合在一起，值得赞赏。

1. Market Street Bridge
2. Riverpool
3. Sundeck
4. Grove
5. Movie Screen and Cinema Lawn
6. Batture
7. Playground and Snack Kiosk
8. Urban Wild
9. Perched Marsh
10. Saint Mary Street Bridge
11. Pavilion
12. Irish Channel Pier
13. Saint Andrew Street Pier
14. Wind Turbines
15. Batture
16. Observation Tower Picnic Lawn
17. Jackson Avenue Ferry Terminal

1. 市场街大桥
2. 河中游泳池
3. 日光浴平台
4. 树林
5. 电影屏幕和影院草坪
6. 河滩
7. 操场和小吃摊
8. 城市荒野
9. 栖息沼泽
10. 圣玛丽街大桥
11. 凉亭
12. 爱尔兰海峡码头
13. 圣安德鲁街码头
14. 风涡轮
15. 河滩
16. 观察塔野餐草坪
17. 杰克逊大道渡轮码头

Notes of Interest

New Orleans has long been dependent on its majestic river. The banks of the Mississippi River have served many purposes throughout the city's history and are now poised to play a crucial new role. The city's economy has suffered the slow loss of maritime activity due to port consolidation and sudden, comprehensive loss of civic stability due to Hurricane Katrina in 2005.

Paradoxically, the hurricane heightened public understanding that the riverfront is in fact the "high ground" and ripe for possible redevelopment. As such, the Reinventing the Crescent Development Plan calls for the East Bank of the city's central riverfront to accommodate a continuous sequence of public open spaces, and along this sequence establish 15 special environments.

Some of these places reinforce and enhance existing public domains, such as improving the riverfront's Moonwalk and creating a better pedestrian connection between the Moonwalk and Jackson Square. Others are new urban nodes allowing the city to reconnect with the river's edge. Each of the new development nodes is strategically located to facilitate the mitigation of physical barriers that have kept citizens at an "urban arm's length" away from their river.

Associate Architects: Hargreaves Associates, Chan Krieger Scieniewicz & TEN Arquitectos
Consultant: James Richardson Economic Consulting, Julie Brown Consulting, Kulkarni Consultants, Moffatt & Nichol, Robinson et al. Public Relations, St. Martin, Brown & Associates
Owner: The New Orleans Building Corporation

合作建筑师：哈格里夫斯事务所、CKS&TEN 事务所
顾问：詹姆斯·理查德森经济咨询公司、
朱莉·布朗咨询公司、
库尔卡尔尼咨询公司、
莫法特 & 尼克、
罗宾森公关公司、
圣马丁·布朗事务所
所有人：新奥尔良建筑公司

Architect / 建筑师
Eskew+Dumez+Ripple
埃斯科+杜麦兹+里波

Location / 项目地点
New Orleans, Louisiana
路易斯安那州，新奥尔良

Photo Credit / 图片版权
© Hargreaves Associates
哈格里夫斯事务所

项目特色

新奥尔良一直依赖于它雄壮的河流。密西西比河两岸在城市的历史中起到了至关重要的作用，现在它则成为了一个具有决定性意义的新角色。由于港口合并和2005年卡特里娜飓风所造成的人口流失，缓慢衰退的海事活动影响了城市的经济。

相反，飓风让公众们意识到了河岸地区的优势和潜在开发的可能性。同样的，新月区开发规划要求城市中央河岸区的东岸设置一系列的开放公共空间，建立15个不同的环境。其中的一些区域提升并强化了现有的公共区域，例如改善了河畔的月亮走道、在月亮走道和杰克森广场之间建立了更好的步行连接。其他区域则成为新的城市节点，让城市与河岸重新联系起来。每个新的开发节点的选址都巧妙地缓和了市民与河流之间的地理障碍。

POLAND FIELDS
POLAND FIELDS BYWATER POINT PORT OF EMBARKATION REDEVELOPMENT
波兰地滨水区出发港再开发工程

PROGRAM / 规划项目
- Concerts & Performances / 音乐会&演出
- Music Festivals / 音乐节
- Community Gardening / 社区园林
- Running / 跑步
- Picnicking / 野餐
- Dog Walking / 遛狗
- Parking / 停车场
- Street car Rides / 有轨电车道
- Lawn Sports / 草坪运动
- Bike Riding / 自行车道

CIRCULATION / 交通
- Japonica Street Bridge Overpass / 山茶街大桥立交桥
- St Claude Avenue Pedestrian Bridge / 圣克劳德大道人行天桥
- Streetcar Extension To Poland Av / 有轨电车被延长至波兰大道
- Neighborhood Access to Poland / 波兰地公园入口通道
- Fields Park / 园地公园
- Cruise Terminal / 游船码头
- Secure Vehicular Zone / 安全车辆区
- Bywater Point Walking Trails / 滨水步行道
- Levee Walk / 大堤步行道

LANDSCAPE / 景观
- Shade Tree Allees / 绿荫树大道
- Streetcar Bosques / 有轨电车灌木丛
- Parking Garden Tree Islands / 停车场花园树岛
- Garden Tree Islands / 花园树岛
- Vegetable and Flower Gardens / 果蔬园和花园
- Cruise Terminal Deck / 游船码头
- Riverview Landforms / 河景地貌
- Amphitheater Lawns / 露天剧场草坪
- Bywater Groves / 滨水树林
- Roof Garden / 屋顶花园
- Existing Batture / 现有河滩
- Levee / 堤坝

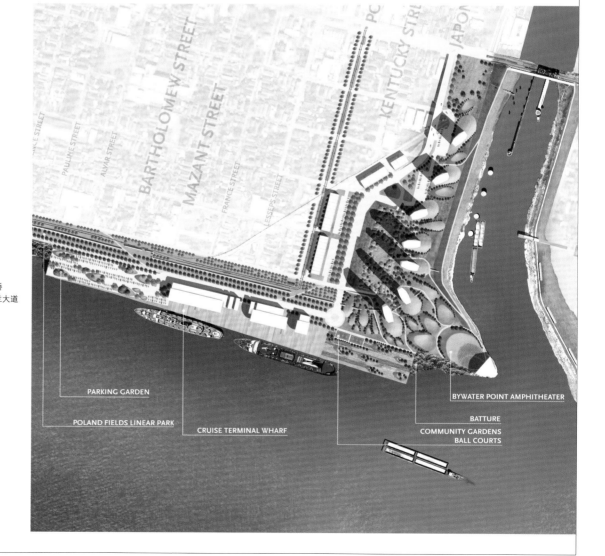

PARKING GARDEN
POLAND FIELDS LINEAR PARK
CRUISE TERMINAL WHARF
BYWATER POINT AMPHITHEATER
BATTURE
COMMUNITY GARDENS
BALL COURTS

SandRidge Energy Commons
沙波能源公司绿地规划

Jury Comments:
This is a particularly refreshing project that takes on a civic role in the redevelopment of existing buildings to create a better downtown.
The reinvestment of this corporate campus project combined with preservation and strong public spaces will contribute to making Oklahoma City a better place.
The park is particularly compelling and the character of the renderings is very strong.

评委评语：
这是一个令人重新振作的项目，在原有建筑的重新开发中起到了公民的角色，打造了一个更好的市中心。
公司园区项目的再投资与环境保护相结合，强健的公共空间将让俄克拉何马市变得更美好。
公园设计引人注目，色彩渲染十分强烈。

SANDRIDGE COMMONS
沙波能源公司大楼公共区

© Rogers Marvel Architects

Notes of Interest

SandRidge Energy, a rapidly growing natural gas and oil company, relocated from the outskirts of Oklahoma City into an abandoned area of the downtown core. The master plan for this new headquarters spans multiple buildings, and multiple city blocks, where architecture and landscape architecture weave to balance company needs and civic engagement.

The project creates a network of programs to support employees while forming a destination location within downtown. The distribution of programs serves as catalysts to encourage development of adjacent properties and integrate the company into the fabric of the city. Shared outdoor spaces enable employees, their families, and the broader community to enjoy spending time downtown.

The project is located between two major arteries – North Robinson, the city's green connector, and North Broadway, lined with commercial buildings and spaces. SandRidge Commons is an "outdoor interior" that provides a green link between these two major arteries.

This community-centered urban design project represents an approach not commonly sought; an approach that holds transferable lessons and great potential for other cities. Rather than becoming an icon and shaping only the skyline, rather than becoming a campus and shaping only an insular world, the task for SandRidge Commons was to weave program, buildings and landscapes, into the urban fabric and help improve the map of the city it calls home.

Construction Management: Lingo Construction Services
Engineer: ARUP; Frankfurt Short Bruza Associates
Landscape Architect: Hoerr Schaudt
Lighting: Microclimate and Daylighting Studies; Renfro Design Group
Owner: SandRidge Energy

施工管理：林果施工服务公司
工程师：ARUP；法兰克福短布鲁扎事务所
景观建筑师：霍尔·沙德特
照明设计：微气候和日光照明研究室；任弗洛设计集团
所有人：沙波能源公司

Architect /建筑师
Rogers Marvel Architects
罗杰斯·马维尔建筑事务所

Location /项目地点
Oklahoma City, Oklahoma
俄克拉荷马州，俄克拉何马市

Photo Credit /图片版权
© Dbox, © Raddi Inc., © Rogers Marvel Architects
Dbox、拉迪公司、罗杰斯·马维尔建筑事务所

AN URBAN SHELTERBELT
城市防护林带

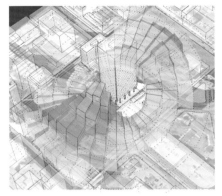

Wind patterns were analyzed to develop mitigation strategies.
研究风势以开发缓解策略。

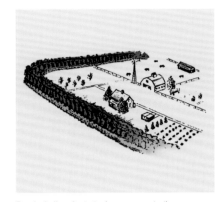

Rural windbreak strategies common to the region were studied to glean the lessons learned from decades of use in the protection of agricultural land.
设计师研究了当地普遍的乡村防风林策略，从农田保护中吸取经验。

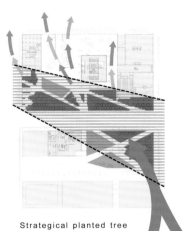

Strategical planted tree windbreaks slow down fast moving wind and shelter new public spaces.
有策略地种植防风树木减缓了风速并保护了公共空间。

© Rogers Marvel Architects

© Rogers Marvel Architects

© Dbox

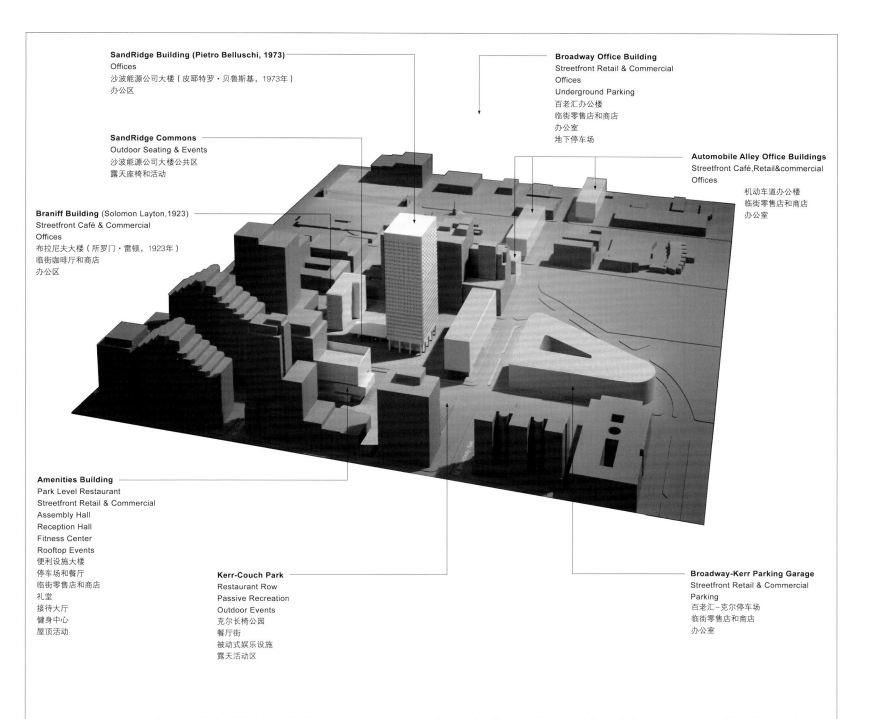

SandRidge Building (Pietro Belluschi, 1973)
Offices
沙波能源公司大楼（皮耶特罗·贝鲁斯基，1973年）
办公区

SandRidge Commons
Outdoor Seating & Events
沙波能源公司大楼公共区
露天座椅和活动

Braniff Building (Solomon Layton, 1923)
Streetfront Café & Commercial
Offices
布拉尼夫大楼（所罗门·雷顿，1923年）
临街咖啡厅和商店
办公区

Amenities Building
Park Level Restaurant
Streetfront Retail & Commercial
Assembly Hall
Reception Hall
Fitness Center
Rooftop Events
便利设施大楼
停车场和餐厅
临街零售店和商店
礼堂
接待大厅
健身中心
屋顶活动

Kerr-Couch Park
Restaurant Row
Passive Recreation
Outdoor Events
克尔长椅公园
餐厅街
被动式娱乐设施
露天活动区

Broadway Office Building
Streetfront Retail & Commercial
Offices
Underground Parking
百老汇办公楼
临街零售店和商店
办公室
地下停车场

Automobile Alley Office Buildings
Streetfront Café, Retail & commercial
Offices
机动车道办公楼
临街零售店和商店
办公室

Broadway-Kerr Parking Garage
Streetfront Retail & Commercial
Parking
百老汇–克尔停车场
临街零售店和商店
办公室

A MORE SUSTAINABLE DISTRIBUTION
更具可持续性的布局

Distributed uses housed within multiple buildings integrate readily with adjacent fabric and offer the ability for rapid change in response to changing economic conditions.
各个建筑内部的布局有效地与周边城市网络结合在一起，适应于经济条件的快速发展。

© Raddi Inc.

项目特色

沙波能源是一家快速发展的天然气和石油公司，他们从俄克拉何马市的郊区移到了市中心一块被废弃的区域。公司新总部的总体规划跨越了多座建筑和多个城市街区，建筑和景观交织在一起，平衡了公司需求和市民参与。

项目打造了一个规划网络来支持员工的日常活动，同时也在市中心形成了一个景点。分散的功能区催化了附近产业的开发并且将公司纳入了城市网格之中。共享户外空间让员工、他们的家庭成员以及更广阔的社区人群在市中心共享时光。

项目位于两条主干道之间——北罗宾逊大道（城市的绿色连接器）和北百老汇大道（两侧林立着商业建筑和空间）。作为一个"露天室内景观"沙波绿地在两条主干道之间建立了绿色连接。

这个以社区为中心的城市规划项目呈现了独特的设计，为其他城市提供了可借鉴的范例。沙波绿地没有成为塑造天际线的地标，也没有成为一个孤立的园区，而是将各个功能区交织在一起，让建筑与景观纳入城市网格，帮助提升了城市的家庭感。

Gehry Residence
盖里住宅

Jury Comments:
Published around the world, the image of a defiantly "destroyed" California house made of unexpectedly humble materials ignited responses as far as Europe and Asia.
As often with ground-breaking efforts, the provocative house invited astonishment, admiration, and contempt. Even with a groundswell of disdain, the house eventually justified its place in architectural history by offering a strong rebuttal to the kitsch neo-historic approach of postmodernism.
It ignited a forum to consider the relationship between art and architecture, which fueled the subsequent waves of architect and artist collaborative projects in the 1980s, further expanding the role of the architect in culture.

评委评语：
项目在世界各地出版发布。这座反叛的"破坏性"加州住宅由不可思议的低调材料建成，在欧洲、乃至亚洲都起到了重大的影响。
正如许多独创性设计一样，这座激进的住宅令人惊奇、赞叹并且轻视。在大量的批评声中，住宅最终在建筑历史上站牢了自己的位置，强烈反驳了对后现代主义的拙劣模仿。
它掀起了人们对艺术与建筑关系的大讨论，促进了随后在20世纪80年代的建筑师与艺术家合作浪潮，并进一步拓展了建筑师在文化中的作用。

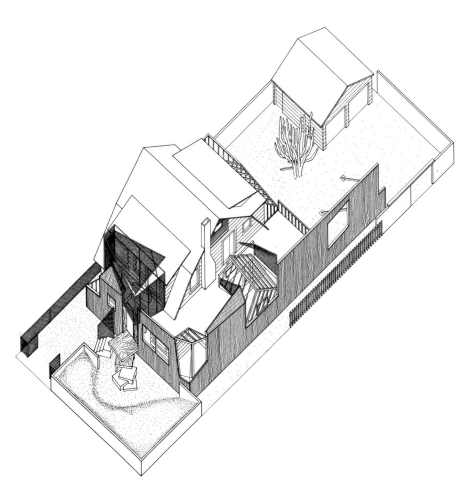

Notes of Interest
A seemingly ad hoc collection of raw, workmanlike materials wrapped around an unassuming two-story clapboard bungalow, Frank Gehry's, FAIA, home for his wife, Berta, and two sons found a literal, but unexpected, answer to the question of neighborhood context, and used it to forever re-shape the formal and material boundaries of architecture.
Enormously influential in both theory and practice, the home's fundamental material modesty and formal experimentation marks a Rubicon in the history of contemporary architecture, tearing down inherited stylistic standbys to declare a new design language for the modern suburban architectural condition. Recognizing architectural design of enduring significance, the Twenty-five Year Award is conferred on a building that has stood the test of time for 25 to 35 years as an embodiment of architectural excellence. Projects must demonstrate excellence in function, in the distinguished execution of its original program, and in the creative aspects of its statement by today's standards. The award will be presented this May at the AIA National Convention in Washington, D.C.

项目特色
盖里住宅看似是一个临时的组合，以未加工的和精巧的材料包围住了一个两层楼高的隔板建筑。弗兰克·盖里为他的妻儿们打造了一座毫不夸张却又出乎意料的住宅，回应了周边的建筑环境，并且能够不断地对其进行造型和材料的改造。
住宅的设计在理论和实践上都具有巨大的影响力。它的低调的基础材料和造型试验在现代建筑历史上写下了浓墨重彩的一笔，瓦解了传统设计，打造了全新的现代城郊建筑典范。25年奖专门授予历经25到35年时间考验的优秀建筑，认可它们持久的影响力。获奖项目必须在功能性、原始规划执行性以及创新性方面具有非凡的表现。本奖项将在2012年5月于华盛顿的美国建筑师协会全国代表大会上颁布。

Architect / 建筑师
Gehry Partners LLP
盖里事务所

Location / 项目地点
Santa Monica, California
加利福尼亚州，圣塔莫尼卡

Photo Credit / 图片版权
© Gehry Partners LLP
盖里事务所

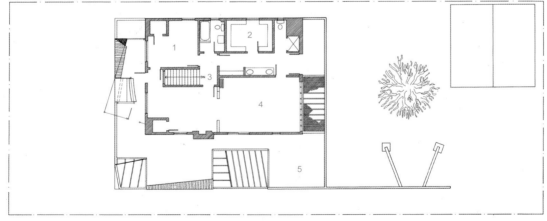

1. Bedroom
2. Closet
3. Down
4. Master Bedroom
5. Outdoor Deck

1. 卧室
2. 壁橱
3. 由此下楼
4. 主卧室
5. 露天平台

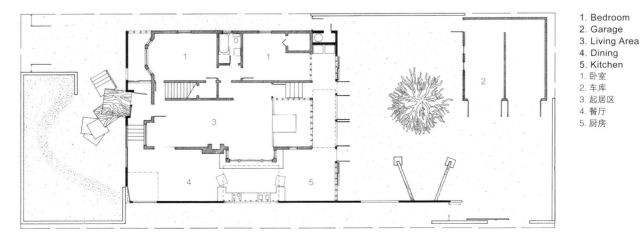

1. Bedroom
2. Garage
3. Living Area
4. Dining
5. Kitchen

1. 卧室
2. 车库
3. 起居区
4. 餐厅
5. 厨房

INDEX
索引

_____2010

Albert & Associates Architects
http://www.albertassociates.com
514 Main Atreet Hattiesburg
Mississippi 390401
Tel: 601.544.1970

Architecture Research Office
http://www.aro.net
170 Varick Street, 7th Floor New York
NY 10013
Tel: 212.675.1870

Barkow Leibinger Architects
http://www.barkowleibinger.com
SchillerstraBe94 D-10625 Berlin
Tel: 49(0)30.315712-0

Bentel & Bentel Architects
http://www.bentelandbentel.com
22 Buckram Road Locust Valley
NY 11560
Tel: 516.676.2880

Burt Hill
650 Smithfield Street
Suite 2600
Pittsburgh, PA 15222
Tel: 412.394.7000

Choi Ropiha
http://chrofi.com/
3/1 The Corso Manly NSW 2095 Australia
Enter from Whistler Street
Tel: 61.2.9977.3700

Conway+Schulte Architects
http://www.conwayandschulte.com
2300 Kennedy Street NE #240
Minneapolis, Minnesota 55413
Tel: 612.333.5867

Dake Wells Architecture
http://dake-wells.com
401 W. Walnut St.
Springfield, Missouri 65806
Tel: 417.831.9904

Daly Genik
http://www.dalygenik.com
1558 10 th St-C
Santa Monica, CA 90401
Tel: 310.656.3180

Diller Scofidio + Renfro
http://www.dsrny.com/
601 W. 26 th Street, Suite 1815
New York, NY 10001
Tel: 212.260.7971

Epstein | Metter Studios
http://www.epsteinglobal.com
600 West Fulton
Chicago, Illinois 60661-1199
Tel: 312.454.9100

FXFOWLE Architects
http://www.fxfowle.com
22 West 19th Street
New York, NY USA 10011
Tel: 212.627.1700

Gabellini Sheppard Associates
http://www.gabellinisheppard.com
665 Broadway Suite 706
New York, NY 10012
Tel: 212.388.1700

Kendall/Heaton Associates, Inc.
http://www.kendall-heaton.com

Kuwabara Payne McKenna Blumberg Architects
http://www.kpmbarchitects.com

Mack Scogin Merrill Elam Architects
http://msmearch.com
111 John Wesley Dobbs Avenue
NE Atlanta, Georgia 30303
Tel: 404.525.6869

Meyer, Scherer & Rockcastle, Ltd.
http://www.msrltd.com
710 South 2nd Street 8th Floor
Minneapolis, MN 55401
Tel: 612.375.0336

Office dA, Inc.
http://www.officeda.com

Olson Kundig Architects
http://www.olsonkundigarchitects.com
159 South Jacjson St.,Suite 600
Seattle, WA 98104,USA
Tel: 206.624.5670

Perkins Eastman
http://www.perkinseastman.com
115 Fifth Avenue
New York, NY 10003
Tel: 212.353.7200

Peter Marino Architect
http://www.petermarinoarchitect.com
150 East 58 Street
New York, NY 10022
Tel: 212.752.5444

PKSB Architects
http://www.pksb.com
330 West 42nd Street
New York, NY 10036
Tel: 212.594.2010

Polshek Partnership Architects
http://ennead.com
320 West 13 th Street
New York,New York 10014
Tel: 212.807.7171

Pugh + Scarpa
http://www.pugh-scarpa.com
4611 W. Slauson Ave.
Los Angeles, California 90043
Tel: 323.596.4700

Randy Brown Architects
http://www.randybrownarchitects.com
1925 N. 120th Street
Omaha, NE 68154
Tel: 402.551.7097

Shelton, Mindel & Associates
http://www.sheltonmindel.com
56 west 22 nd street,12 th floor
New York, NY 10010
Tel: 212.206.6406

Skidmore, Owings & Merrill LLP
http://www.som.com
224 S. Michigan Avenue Suite 1000
Chicago, IL 60604, USA
Tel: 312.554.9090

Sottile & Sottile
http://www.sottile.cc

Thomas Phifer and Partners
http://www.tphifer.com
180 Varick Street
New York, NY 10014
Tel : 212.337.0334

Tod Williams Billie Tsien Architects
http://www.twbta.com
222 Central Park South
New York, NY 10019
Tel: 212.582.2385

Wallace Roberts & Todd
http://www.wrtdesign.com

WSA Studio
http://wsastudio.com
The Jack, 982South Front Street
Columbus, Ohio 43206
Tel: 614.824.1633

2011
Adrian Smith + Gordon Gill Architecture
http://smithgill.com
111 West Monroe, Suite 2300
Chicago IL 60603
Tel: 312.920.1888

Allied Works Architecture
http://www.alliedworks.com
12 W 27th St, 18th Floor
New York, NY 10001
Tel: 212.431.9476

Belzberg Architects
http://www.belzbergarchitects.com
2919 1/2 Main Street
Santa Monica, CA 90405
Tel: 310.453.9611

Bernard Tschumi Architects
http://www.tschumi.com
227 West 17th Street, second floor
New York, New York 10011
Tel: 212.807.6340

Cannon Design
http://cannondesign.com
100 Cambridge Street, Suite 1400
Boston, Massachusetts 02114
Tel: 617.742.5440

Clive Wilkinson Architects
http://clivewilkinson.com
Los Angeles
144 North Robertson Boulevard
West Hollywood, CA 90048
Tel: 310.358.2200

dlandstudio llc
http://www.dlandstudio.com
137 Clinton Street
Brooklyn, NY 11201
Tel: 718.624.0244

Lehrer Architects
http://lehrerarchitects.com
2140 Hyperion Ave
Los Angeles, CA 90027-4708
Tel: 323.664.4747

Jensen Architects/Jensen & Macy Architects
http://www.jensen-architects.com
833 Market Street, 7th Floor
San Francisco, CA 94103-1827
Tel: 415.348.9650

Julie Snow Architects, Inc.
http://www.juliesnowarchitects.com
2400 Rand Tower
527 Marquette Avenue
Minneapolis, Minnesota 55402
Tel: 612.359.9430

KlingStubbins
http://www.klingstubbins.com
Washington, DC
2000 L Street, NW, Suite 215
Washington DC 20036
Tel: 202.785.5800

Kohn Pedersen Fox Associates, PC
http://www.kpf.com
11West 42nd Street
New York, NY 10036
Tel: 212.977.6500

Lake | Flato Architects
http://lakeflato.com
311 THIRD STREET
San Antonio, TX 78205
Tel: 210.227.3335

LMN + DA/MCM
http://lmnarchitects.com
801 Second Avenue, Suite 501
Seattle, Washington 98104
Tel: 206.682.3460

Marcy Wong Donn Logan Architects
http://wonglogan.com
800 Bancroft Way Suite 200
Berkeley, CA 94710
Tel: 510.843.0916

Marilys R. Nepomechie Architect
Associate Professor of Architecture at Florida International University

Marta Canaves Interior Design
7373 Sw 60th Street
Miami, FL, 33143

Montalba Architects, Inc.
http://montalbaarchitects.com
2525 Michigan Avenus, Bldg.,T4
Santa Monica, CA, 90404
Tel: 310.828.1162

Patrick Tighe Architecture
http://www.tighearchitecture.com
1632 Ocean Park Blvd
Santa Monica, CA 90405
Tel: 310.450.8823

Pei Cobb Freed & Partners Architects LLP
http://www.pcf-p.com
88 Pine Street
New York, NY 10005
Tel: 212.751.3122

Rene Gonzalez Architect
http://renegonzalezarchitect.com
670 NE 50th Terrace
Miami, FL 33137-3023
Tel: 305.762.5895

REX | OMA
http://www.rex-ny.com
20 Jay Street Suite 920
Brooklyn, NY 11201
Tel: 646.230.6557

Skidmore, Owings & Merrill LLP
http://www.som.com
224 S. Michigan Avenue, Suite 1000
Chicago, IL 60604
Tel: 312.554.9090

Steven Holl Architects
http://www.stevenholl.com
New York City
450 West 31st Street, 11th floor
New York, NY 10001
Tel: 212.629.7262

Thomas Phifer and Partners
http://www.tphifer.com
180 Varick Street
New York, NY 10014
Tel: 212.337.0334

University of Arkansas Community Design Center
http://uacdc.uark.edu
104 N. East Avenue
Fayetteville, AR 72701
Tel: 479.575.5772

Weiss/Manfredi Architecture/Landscape/Urbanism
http://www.weissmanfredi.com
200 Hudson Street 10fl
New York, NY 10013
Tel: 212.760.9002

ZGF Architects LLP
http://www.zgf.com
Portland Office
1223 SW Washington Street, Suite 200
Portland, Oregon 97205
Tel: 503.224.3860

_____2012

Andre Kikoski Architect, PLLC
http://www.akarch.com
180 Varick Street,Suite 1316
New York, NY 10014
Tel: 212.627.0240

Anmahian Winton Architects
http://aw-arch.com
650 Cambridge Street
Cambridge, MA 02141
Tel: 617.577.7400

BIG
http://www.big.dk
BIG NYC
601 West 26th Street, Suite 1255
New York, NY 10001
Tel: 347.549.4141

Center for Design Research, School of Architecture + Design, Virginia Tech
http://www.lumenhaus.com
201 Cowgill Hall
Blacksburg, VA 24060
Tel: 540.818.5012

Cooper, Robertson & Partners;
http://www.cooperrobertson.com/
311 West 43rd Street
New York, NY 10036
Tel: 212.247.1717

David Jameson Architect
http://www.davidjamesonarchitect.com
113 South Patrick Street
Alexandria, Virginia 22314
Tel: 703.739.3840

DMOD Architects
http://www.dmod.ie
Cathedral Court, New Street
Dublin 8 Ireland
Tel: 353.1.491.1700

Ennead Architects
http://ennead.com
Ennead Architects LLP
320 West 13 th Street
New York, NY 10014
Tel: 212.807.7171

Eskew+Dumez+Ripple
http://www.eskewdumezripple.com
one canal place
365 Canal Street Suite 3150
New Orleans, LA 70130
Tel: 504.561.8686

Gehry Partners LLP
http://www.foga.com
12541 Beatrice Street
Los Angeles, CA 90066
Tel: 310.482.3000

Gensler
http://www.gensler.com
Rockefeller Center
1230 Avenue of the Americas,
Suite 1500
New York, NY 10020
Tel: 212.492.1400

Goettsch Partners
http://www.gpchicago.com
224 South Michigan Avenue, Floor 17
Chicago, Illinois 60604
Tel: 312.356.0600

John Ronan Architects
http://www.jrarch.com
420 W Huron Street
Chicago, Illinois 60654
Tel: 312.951.6600

KieranTimberlake
http://kierantimberlake.com
420 North 20th Street
Philadelphia, PA 19130.3828
Tel: 215 922 6600

Koning Eizenberg Architecture
http://www.kearch.com
1454 25th Street
Santa Monica, CA 90404
Tel: 310.828.6131

Mack Scogin Merrill Elam Architects
http://msmearch.com
111 John Wesley Dobbs Avenue, NE
Atlanta, Georgia 30303
Tel: 404.525.6869

Mackay-Lyons Sweetapple Architects Limited
http://www.mlsarchitects.ca/portfolio/featuredprojects
2488 Gottingen Street
Halifax, Nova Scotia Canada B3K 3BK 3B4
Tel: 902.429.1867

Marlon Blackwell Architect
http://www.marlonblackwell.com
The Fulbright Building
217 E. Dickson St., Suite 104
Fayetteville, Arkansas 72701
Tel: 479.973.9121

Michael Maltzan Architecture, Inc.
http://www.mmaltzan.com
2801 Hyperion Avenue, Studio 107
Los Angeles, California 90027
Tel: 323.913.3098

Moore Ruble Yudell Architects & Planners;
http://www.mryarchitects.com

Morphosis Architects
http://www.morphosis.com
3440 Wesley Street
Culver City, CA 90232
Tel: 424.258.6200

OLIN
http://www.theolinstudio.com
Public Ledger Building, Suite 1123
150 South Independence Mall West,
Philadelphia, PA 19106
Tel: 215.440.0030

Patrick Tighe Architecture
http://www.tighearchitecture.com
1632 Ocean Park Blvd
Senta Monica, CA 90405
Tel : 310.450.8823

Randy Brown Architects
http://www.randybrownarchitects.com
1925 N. 120th Street
Omaha, NE 68154
Tel: 402.551.7097

Rogers Marvel Architects
http://www.rogersmarvel.com
145 Hudson Street, Third Floor
New York, NY 10013
Tel: 212.941.6718

Sasaki Associates, Inc.
http://www.sasaki.com
64 Pleasant Street
Watertown, MA 02472
Tel: 617.926.3300

Shim-Sutcliffe Architects
http://www.shim-sutcliffe.com
441 Queen Street East
Toronto, Ontario, Canada M5A 1T5
Tel: 416.368.3892

Tod Williams Billie Tsien Architects
http://www.twbta.com
222 Central Park South
New York, NY 10019
Tel: 212.582.2385

University of Arkansas Community Design Center
http://uacdc.uark.edu
104 N. East Avenue
Fayetteville, AR72701
Tel: 479.575.5772

West 8 Urban Design and Landscape Architecture
http://www.west8.nl
Schiehaven 13M, 3024 EC
Rotterdam, the Netherlands
Tel: 31(0)10.485.5801

ZGF Architects LLP
http://www.zgf.com
Portland Office
1223 SW Washington Street, Suite 200
Portland, Oregon 97205
Tel: 503.224.3860